What Readers Say About BUILD IT RIGHT!

Out last home was built before we read *Build It Right*, our next home was planned after having read the book. What a difference! We thought we knew all the trick little details to watch out for in a house. After reading *Build It Right*, we realized how many things we had missed.

Tom and Lisa Howard, Salem, OR

We got *Build It Right* too late—after we had committed to our new tract home. Fortunately there were some things that could still be changed. It would have been a better house if we had had the book earlier.

Chris and Cindy Pflaumer, San Jose, CA

I found *Built It Right* an informative and helpful reference while having our home designed. We had an excellent architect and he had avoided most of the pitfalls—still it was good to have a list for double checking!

Patricia M. Ezell, Hernando, FL

Build It Right was a tremendous "guide book" in making the multitude of decisions on building our new home. One suggestion I found very helpful was window and door placements which would allow for adequate wallspace for the furniture I had. We incorporated several pocket doors. Thanks!

Elaine V. Graeff, Marshalltown, IA

When I set out to look for a new home I was surprised at how many of the design flaws discussed in *Build It Right* I saw in models. It made me a much more informed and critical shopper.

M. J. Stackowski, MD, Milton, PA

I read *Build It Right* from cover to cover and then went over the plans for my house. I changed a closet door that would have been covered by the door into the room. I shortened the width of a wall next to the refrigerator and even changed from a side by side to an up and down refrigerator as a result of reading the book.

Kathleen Guy, Rotonda West, FL

We were very impressed with the suggestions we found in *Build It Right*—the space for oven doors to open as well as dishwashers. Placement of light fixtures was very helpful to mention just a few.

Veronica L. Hess, Novi, MI

What Reviewers Say

Myron Ferguson has studied details in his own homes and in more than 700 other new houses...The product of this research is *Build It Right!*, a book that is as interesting for its anecdotes as much as for its advice...cautionary tales are woven through chapters on selecting a lot, house designs, whole-house systems, the kitchen, bedrooms and baths, and finding and working with a builder. The...book includes diagrams and photos that illustrate significant points.

> Robert Wilson, Editor, *Better Homes and Gardens Building Ideas*

If you're building a new home or remodeling the old homestead, *Build It Right!* will help you plan and complete the project with maximum results and minimum headaches. Myron Ferguson has given the consumer a great tool to use in navigating the pitfalls of the design and construction jungle.

> Jordan Clark, President, United Homeowners Association

New-house how to...If you're gathering material about...a house, you might add "Build It Right!" to your collection. It's by Myron E. Ferguson, a systems engineer, who was dissatisfied with his own new home.

He takes the reader through one spot at a time—how to plan your attic if you want to use it for storage, how to avoid doors that swing awkwardly.

> Judy Rose, Homes Editor, *Detroit Free Press*

This Book Is a Must for You...He (Ferguson) provides hundreds of tips on design and building materials...He explains how to assess existing floor plans and how to modify them, how to work with an architect and how to find a competent builder.

> Around the House, Les Hausner, *Chicago Sun-Times*

If you're thinking of building a home, get set to ask lots of questions and make lots of decisions. *Build It Right*...is full of detailed yet easy-to-understand information about kitchens, beds and baths, electrical and plumbing systems, windows, exteriors and more. Includes photos, glossary...recommended reading.

> Judy Stark, Homes Editor, *The St. Petersburg Times*

BUILD IT RIGHT!

What to Look for in Your New Home

Revised Edition

MYRON E. FERGUSON

HUP
HOME USER PRESS

HOME USER PRESS
SALEM, OREGON

HOME USER PRESS
1939 Woodhaven Street NW
Salem, OR 97304-1854

Revised edition, first printing

Publishers Cataloging-in-Publishing
(Provided by Quality Books, Inc.)

Ferguson, Myron
     Build it right! : what to look for in your new home / Myron E.
Ferguson. -- Rev. ed.
     p. cm.
     Includes bibliographical references and index.
     ISBN 0-9654856-0-9
     Preassigned LCCN: 97-093290

     1. Buildings—Defects.   2. House buying.   I Title
     TH441.F47 1997   643'.12'0979
                                       QBI97-40368

This book is a work of advice and opinion. Neither the author nor the
publisher is responsible for actions based on the contents of this book.

Edited by Linda West
Cover design by RL Graphics
Cover illustration courtesy of Alan Mascord Design Associates

Foreword

You have decided to buy a new home—a place of your own—a statement of who you are. You are probably excited and a bit nervous at the prospect. Myron Ferguson's goal in writing *Build It Right!* is to take much of the confusion and uncertainty out of the process. Whatever your concerns about getting a new home, *Build It Right!* should help you feel more confidant and self assured—because you are undertaking one of the most important projects of your life.

Build It Right! takes you step by step through the home building process as your dream becomes a reality. One of the greatest challenges in getting that new house is understanding what is "right" for you. Reading Mr. Ferguson's book will help you make those determinations before you move in and find that the light switch is in the wrong place or the closet is too small. *Build It Right!* helps you not only define "quality" but helps you understand how to communicate that definition to your builder.

Although most jurisdictions operate with a building code that establishes standards for design and construction, these codes were generally developed with safety and health in mind. They often vary from city to city and cover aspects of the home in different ways. This obviously leads to confusion in what is the "right" way to build it. As a result, it is extremely difficult to define acceptable quality in relation to materials and workmanship—this often leaves customers with differing expectations for the definition of "quality" as it relates to the finished product. In his book *Build It Right!* Myron Ferguson seeks to design a type of reference manual for the would-be educated home buyer and answer questions pertaining to quality of construction. This revised edition should help you communicate your home building vision, be realistic about what you can afford, and better understand how you will use your new home.

Happy, satisfied customers are the backbone of the home building industry. The better that customers are able to express their expectations, the better the builder will be able to meet them. BUILD IT RIGHT! develops a platform so expectations can be dreamed and developed by you—then diagnosed, designed, and delivered by your builder.

Best wishes on a rewarding, exciting adventure.

Jim Irvine

Jim Irvine is a Portland, Oregon home builder, was the 1995 president of the National Association of Home Builders, is active in housing issues at local, state and national levels, and currently chairs the Pacific Northwest Delegation to the White House Conference of Small Businesses.

Preface

This revision of *BUILD IT RIGHT!* is the culmination of five years of research, three years of it since the first edition was written. I have looked at about 1,000 new homes, on both coasts of our country, and have had the benefit of interesting and informative discussions with people who bought the first edition at home shows and ordered it on the telephone.

Even before the first edition was published I was well aware that much of the information in the book was just as applicable to tract and spec homes as to custom homes. I received a number of comments to the effect that there should be a book for these other buyers of new homes. At the same time, the feedback I was getting was that the parts of the book about basic designs were the most useful.

These inputs led to this revised edition where the stress is on what makes a home a better place to live—regardless of who designs or builds it and regardless of whether it is in a tract, is a spec house, or is one that is customized.

When trying to explain the user-friendly and user-hostile things I've seen in new homes, I always feel more comfortable when I can show what I mean visually. This led me to significantly increase the number of illustrations in this revision. There are 68 photos and over 100 drawings. This combination of words and graphics make it easier for the reader to understand the points that are made.

A large proportion of the houses I've looked at were in tracts made both by large national companies and by smaller local builders. They ranged from starter homes to large expensive ones. The problems I saw, however, were universal. The photographs will help the tract or spec buyer recognize the problems when he/she is looking at models and the drawings will make it easier for the custom home buyer to see the problems when they are on paper.

I am well pleased with how the first edition was received. The only sour note was by the editor of a national magazine whose readers are custom home builders. And his complaint was that I had portrayed builders unfairly. He should be pleased with this revision. Among the things I have come to realize in these last three years is that builders simply reflect what their customers are thinking and, in fact, if they don't, they won't stay in business very long. So the emphasis here is trying to get the readers to be actively involved in the decisions that are made about their new homes. When people don't buy user-hostile houses, they won't be built.

As I continue to look at new houses I am still occasionally surprised to see things that are new to me—both friendly and unfriendly. By the time you read this I will have started a new collection of notes and photos. These will become available as a supplement and eventually as part of a future edition. The more data that people buying new homes have available to them, the better the houses will be that they buy. And this, in turn, will lead to better homes being designed and built.

I can use your help, too. If you run across something that you feel is particularly good or bad, tell me about it (telephone, FAX, and addresses below). You'll have my thanks and the thanks of others who follow you in the home-buying process.

To complete the data that should be in a preface: I retired from a career as a communications system engineer before moving to Oregon and buying a new home in 1991. The Oregon house, a spec house, was the trigger that made me take a closer look at what builders were making and selling. After looking around I decided that ours wasn't as bad as many others. I also decided that I'd turn my engineer's way of thinking to good advantage by helping people get more user-friendly homes. Which led to BUILD IT RIGHT!

I took all of the photos and made all of the drawings. Linda West did the text editing.

My particular thanks go to Dick and Mary Lutz of DIMI Press who published the first edition of BUILD IT RIGHT! and who encouraged and helped us set up Home User Press to publish the revision as well as future books for home buyers.

And, of course, special recognition is due to Jean—my better half—whose thoughts and patience have been an essential part of getting this book into your hands.

Myron E. Ferguson
Home User Press   (TEL) 503/391-8106
1939 Woodhaven Street NW   (FAX) 503/375-2939
Salem, OR 97304-1854   E-mail: mefhup@open.org

Contents

Part I

Realizing the Dream

Part II

The Systems

Part III

The Kitchen

Part IV

Beyond the Kitchen

Part V

The Appendices

On one of our trips Jean and I had dinner at Ray's Boathouse, a picturesque Seattle restaurant set on Puget Sound with a view of the Olympic mountains. We watched the lights of the boats coming into the ship canal until our dinner was served. We commented to Jennifer, the server, about the setting and how attractively the food was presented. She thanked us and then added, "but wait until you taste it." She was right; it was even better than it looked.

We couldn't help but think about the parallel between her comments and a new home. It should be attractive and in a nice setting but once you get past the looks there needs to be something for you to appreciate and enjoy or the looks don't mean a thing. **Bon appétit.**

Introduction

"Houses are built to live in, not to look on; therefore, let use be preferred before uniformity, except where both may be had," *wrote philosopher Sir Francis Bacon almost 400 years ago.*

Then in the late 1800s noted architect Louis Sullivan said the same thing with his "IDEA" that "Form Follows Function."

The years go by—but whether it's 1600, 1900, or 2000, some things simply don't change.

Are today's houses the best available for the money? As we will see, most definitely not! As long as more thought is given to aesthetics and little, if any, to function, people will buy what they perceive is attractive. So that's what's built. But why shouldn't you get a house to, as Bacon said, "live in" as well as "look on?" You should and you can.

Welcome to the world of houses and homes, a world as imprecise and varied as the people who live in them. Nowhere does our individuality show more than in our choice of new homes. We have this dream of something that is different from anyone else's, something worthwhile that we personally are responsible for, something that we work for and can be proud to say is ours.

The idea that a new home is the "American Dream" may be a marketing stratagem, yet it has an underlying truth that we all understand. This dream is an edifice that one stands and admires—the flow of the roof, the beckoning of the entry, the magnificence of the whole. Inside, the furnishings are also of dream stuff, making day-to-day living luxurious and relaxing.

The first struggle in making this dream a reality is, for most of us, simply money. Instead of the wooded estate, we know we'll have to settle for the cottage in a neighborhood of peers. Yet we can't forget that dream so we compromise here and take less there, making sure we express ourselves in the process. And we find lots of people who want to help us along the way—people who make a living at helping the dreamers into the real world. They show us what others have done, they find out about our dreams, then work with us to get the best we can with the resources we have.

These helpers we know as real estate people, designers, architects, and builders. They have experience in the many facets of new homes, experience that we must draw on to avoid disaster. Without them our dreams would never leave the realm of the ethereal. And, of course, we know that people are not perfect so we try to find the helpers who will be the best for us and our goals. They will be our advisers as well as the suppliers of goods and services.

We try, yet it doesn't always work out as well as we had hoped. Sometimes we find that we simply cannot have what we want with the money we have. Or there are practicalities like building codes that keep us from getting there. And still other times, and this is the most painful, when we're through with the whole process, we find that our dream isn't as dreamy as we had expected.

The professionals all help us get what we want, but what happens when we don't know what we should want? We learn the hard way—whence come the expressions "if we had only thought of it before..." and "if we had it to do over again..."

Designers create houses they believe are what people want, builders build them and consumers, with their limited experience, buy them. Of necessity, we consumers depend on the designers, builders, and sellers to give us the best we can get for our money. But these professionals, in their efforts to please, produce what people want which is what people buy which is what is produced... From Bacon to Sullivan to today, it's the same story.

Houses are complex structures made of many different materials from many sources. The suppliers of these materials are large industries in themselves: lumber, windows, doors, plumbing, appliances, etc. It is the home-building industry that puts these together into marketable products that are attractive and functional. If there is one thing that characterizes this industry, it is the number and diversity of builders. Unlike other major industries in today's world, there is no big name, no "big three." Most builders do only a few houses a year and even the largest builder (Centex) makes less than two percent of the country's total number of new houses

Given the large number of builders you shouldn't be surprised to find all kinds of people in the business of home construction. As with any profession, there will be the able and the inept, those who take pride in their work and others who don't care, those who build with the customer in mind and those whose end-all is the almighty dollar, those who are honest and forthright and others who are litigious and a disgrace to their industry. They're all out there.

When you buy a new car, there is a large but finite number of options; in new homes the choices are unlimited. From the new home buyer's perspective, this is both good and bad. There is the opportunity to get exactly what you want and can afford. With this freedom goes the flip side, you better know what you want, what's right and what's wrong for you.

Home building is a complex process involving many disciplines and, ideally, the builder would know every one of them intimately. Practically they won't but one discipline is of overriding importance. Builders are small- to medium-sized businesses and they *must* understand the business side of what they're doing or they invite disaster. Unfortunately those disasters usually involve everyone around them: suppliers, subcontractors, *and* customers.

As a manager, the builder should know good and not-so-good materials and workmanship. Scheduling and coordinating material deliveries and subcontractor work are critical and oversight during construction is crucial. Inept and don't-care builders turn things over to the subcontractors and accept whatever happens while good builders stay right on top of the work, with frequent visits to the site to insure coordination and to get boo-boos fixed before they get locked into the fabric of the house.

If the end product is to be acceptable to the customer, the builder must know what's to be built. If it's a custom house, the customer tells the builder what's wanted. If it's a tract or spec house, the question is one of marketability and the builder makes what his research tells him is most likely to sell. The common thread here is that the builder is responding to his/her conception of the marketplace. And as long as buyers don't insist on functionality, along with aesthetics and structural soundness, then functionality will suffer.

And where does all this leave you, the prospective new-home buyer? The first flag should be the necessity to find a competent, honest, caring builder. And this is as true for tract and spec houses as for custom. The second flag is the need for you to know as much as you can about what you want—not only in the size and location of the house but also in its design and workmanship. And, most importantly, being able to make functionality a part of the design.

While we won't change the way the housing industry does its business, you can sure do something about your own new home! In most cases there is little conflict between form and function, between aesthetics and how it works. Rather, *it's simply a matter of remembering function*. And it's not difficult. None of the facets of home design and construction discussed in this book are complicated in themselves. As you read about them you'll say, "Now that makes sense" and you're on your way to a better home.

And it doesn't matter whether your new home is one that you're going to have built to your specifications (a "custom" home), one that's more or less a copy of an existing model (a "tract" home or a "semi-custom" home), or one that already exists (a "spec" home). In the first case you'll want to know what to have and not have in your specifications. For the others you'll want to be able to evaluate models to decide which is the best one for you. And after you decide, you'll need to know which of the available options to choose and what you should try to have the builder do that's different from the model. Many of the decisions you should make are the same regardless of which road you take to get your home—in fact about the only differences are that with a tract house there will be some builder's decisions you cannot change while with an existing spec house you can't change any of them.

The many decisions which have to be made in getting a new home will inevitably involve compromise. You'll need to prioritize your objectives, to decide which of the many things you'd like to have are the more important and which are expendable. Cost is the biggest factor but it isn't the only one. While the stress in BUILD IT RIGHT! is on functionality, don't hesitate to express yourself by making something look better to you even if you compromise functionality in the process. But make these decisions knowingly; don't just let them happen. It is one of the purposes of BUILD IT RIGHT! to help you understand the pros and cons of each compromise and so make your new home reflect *your* personal likes and dislikes.

This revised edition has been completely reorganized. Subject matter has been made more visual—there are sixty eight photographs instead of thirteen and a hundred and nine drawings instead of forty one. Thirty two pages have been added. The scope has been expanded in a number of areas to make the discussions more comprehensive while some less important items have been deleted to make room. The book has been broadened to make it useful to anyone buying a house. The reader is reminded again and again that the subjects covered in the book are not routinely considered in home design and construction—it's very much up to you if they are to be a part of your new home.

Every time I talk to someone about their new home I am reminded again about how many decisions must be made. The book couldn't possible cover all of them because everyone's situation is different—and we wouldn't really want it otherwise. On the way to making your dream a reality you'll have to make a number of compromises usually involving costs and personal likes and dislikes. But availability of materials and the skills of the tradespeople can also make you veer from what you'd really like. With BUILD IT RIGHT! you'll have a better understanding of what you're getting and giving up.

The material in BUILD IT RIGHT! isn't sensational. It won't make *60 Minutes* or *20/20*. It doesn't deal with what happens in a few dramatic disastrous cases but with what happens day in and day out in house after house—in other words with what's likely to happen in your new home if you let it. And what's nice is that you don't have to be a graduate engineer or an interior designer to take advantage of the information that's here. You'll find that most of it is just a matter of having it pointed out to you and you can take it from there.

Part I of the book starts with an overview of what's involved in getting that dream home—tract, spec or custom. The second part deals with the systems that are a part of the whole house and with pieces and parts that are found in every room. Part III is about the kitchen, that most talked about and most abused room in the home. Part IV is about the rest of the rooms: bed, bath, etc. Part V has two appendices that are broad in nature and do not deal directly with the building process.

Finally, there is the back matter—a glossary, a list of references for additional information, and the index.

In 1994 I reviewed 697 floor plans that I had collected from many different sources—almost a third of them from custom and tract houses I had been through. I looked for 9 different design flaws in each plan to get a statistical feel about how often they appeared.

The frequency of occurrence is mentioned where each flaw is discussed in the book. As you will see, user-hostile designs are common.

A word about *Caveat Emptor* (or Buyer Beware). This is used in the book to draw attention to the importance of a particular subject. It is not meant to indicate that someone may be trying to cheat or take advantage but to warn you to beware of the subject at hand; the perils may not be self-evident.

And, before we're through here, there's another matter which must be covered even though it may seem out of place: gender. To simplify the writing and the reading of the book, builders are referred to as "he" rather than "he/she" and is in no way meant to downplay the ever-increasing number and importance of women builders. From my observations women bring a somewhat different set of priorities to home building than do men, particularly in making homes more user-friendly. Regardless of efforts to make the workplace, and even the home, asexual, we still raise our girls to be women and boys to be men. And in the process women learn a different set of values which have been generally ignored in home design and construction. When I've heard "it was built by men" in reference to a particularly unthinking arrangement, it was usually a pertinent observation because it was indeed built by men and it was something a woman builder would be much less likely to do. You'll find more than one of these described in the book.

The concept that homes are not necessarily designed and built for the people who use them is not new—and, generally, it's not likely to change much. But you *can* do something about yours. It'll take work and perseverance. Good luck.

If you get the opportunity to read Louis H. Sullivan's, "The Autobiography of An Idea," you will find it an interesting commentary on the architectural practices of the late 19th century, the practices that led to his "Form Follows Function." The book is long since out of print (it was first published in 1924). But you may find it, as I did, in your local library. It is an interesting way to spend a few hours.

Part I

Realizing the Dream

This part of the book is your introduction to the people who can change your dream into reality—the people who do the work of getting a home designed and built. In many ways the design is even more important than the building. Good designs should result in a home that pleases you; bad designs cannot be fixed by the best builder in the world.

Our object here, as in the rest of *BUILD IT RIGHT!*, is to help you get the best house you can. When you know the things that make a house a better place to live, you'll have a home that's a cut above the norm.

The first chapter is the "big picture," an overview of what's involved in getting a new home, whether tract, spec, or custom.

It always helps a buyer to know something about the organizations with whom he is dealing. The second chapter is a summary review of the trades and professions involved in designing, building and selling houses. It also notes the importance of the code enforcement officer.

Critical first steps are deciding on a lot and choosing a design. These are the subjects of the last two chapters in Part 1.

Chapter 1

From Dream to Reality

Y ou've put in a long day at work and you come home to relax and put the day behind you. What you don't need is a house that makes you or your spouse take extra steps to fix dinner or that reminds you how far it is from the garage to the kitchen with what you've picked up at the store. What you **do** need is a home that doesn't have the irritations and inconsiderations that characterize today's houses. What you need is a user-friendly home.

User-Friendly Homes

User-friendly is a key characteristic of computers—both hardware and software—that says they were deliberately designed to make them easier for users to do a better, faster job. Computers weren't always that way. Twenty years ago the emphasis was on making something that would get a job done and if it was a long and tedious task for the operator it was OK. That has changed.

But not in our homes. Materials today are better than our parents had in theirs and concerns with energy have made houses more comfortable to live in. Yet new homes are no more user-friendly than theirs. Take a look at floor plans from ten, fifteen, or more years ago and you'll find that, as with women's fashions, the "in" things come and go, but there is no more emphasis on utility now than then. Most often aesthetics come before function—if function is thought of at all.

A dictionary definition of value is "the desirability or worth of a thing; intrinsic worth; utility." User-friendly houses have value. The utility or usefulness of a house is directly related to how well its design and construction reflect the needs of the people who live in it.

In the following chapters some of the things we discuss would cost nothing to do the user-friendly way. Many more are low-cost items you'd be happy to pay for if you were given the chance. Being sure your house is user-friendly is a problem for the buyer simply because designers and builders don't think that way. They're in business to sell plans and build houses and, most often, they'll give you what you want. Which, of course, doesn't keep you out of trouble at all.

What's in this book will help you make decisions from a position of knowledge rather than ignorance or apathy.

Resale

When we buy, we may have no intention to move again. Yet we know that's not likely to be the case. Ours is a mobile society; most of us move after living in a house for less than 10 years. (Why do you think there are so many real estate agents?) Eventual resale is something you should keep in mind right from the start.

If you invest in a car, you're concerned about its resale value. If you invest in the stock market, the resale value of the stock is of the utmost importance. Yet when you consider an investment that will take a big hunk of your income for as long as you own the house, how much time do you spend in considering its resale value? The importance of this will come home to roost when you decide it is time to move and suddenly find that, while your house is your dream, it's not anyone else's. Give a little thought to this before you buy or build; it can ease a lot of pain later. As you read, you'll find little reminders about the investment side of your new home.

Regional Preferences

Closely tied to the resale value of your house are the anomalies of regional preferences. These are things that are done one way in one place and differently in another. In some cases these differences came about due to climatic variations between regions. In others, and this is the usual case, the differences exist because that's how it's been done for years and years. From a builder's perspective it is important to follow these regional preferences because there could be trouble selling homes that are different from what people expect to see.

A state building code official had another word for it: tradition-alism. Call it what you want, these differences are not always to your advantage. In Arizona, for example, the shelf in an under-the-counter cabinet typically starts 6 to 8 inches back from the front of the cabinet, noticeably reducing the amount of storage. And in the mid-Atlantic states expect to see the front of this shelf a full 12 inches back! In California, you'll have to special order pull-out shelves in your lower cabinets, because they're not included as a rule, even in upscale houses. In central Pennsylvania stainless steel is preferred to cast-iron for kitchen sinks while in western Kentucky expect ceramic tile as your kitchen floor. And in some areas (the Pacific Northwest) don't look for medicine cabinets in bath-rooms—they're just not there. The list goes on and on.

In some cases, you may choose to not follow the local custom and never be sorry. In others, you may be sorry you bucked the local way of doing things. Be careful to distinguish between doing something that's simply unusual and that which clashes with local tradition. When you get too far off the beaten path, it may mean a tough resale when the time comes. (On the other hand, a house without the user-hostile local customs may be a welcome change for many buyers.)

Your Home Buying Options

When you decide to buy a new home, one of your early decisions will be whether to buy a house that has already been built (a spec house), to buy a house that is a copy or a modified copy of a model in a tract, or to have your dream home custom built.

Many prospective home-owners believe that custom houses cost more than spec or tract houses. This isn't necessarily so. Most often, the extra cost of getting a custom house is not as much in dollars as in your involvement in the process.

Spec Houses

You can, of course, buy an existing spec house where any changes you want will be done remodelling—and all on your nickel. Getting to know what's in a user-friendly home—and what's not in it—should be a prerequisite before you make any decisions. Then, when you start looking at existing houses, you are much better equipped to recognize features that won't be a pain in the neck after you move in.

You may opt to buy the pre-built house anyway, either because you found one you can live with or because the price is so good that you just can't turn it down. What you are more apt to do is to realize that your best chance of getting that better home is to get one that hasn't yet been built, tract or custom, where you have more choice in what it is and what goes into it.

Tracts

Tract houses are easier than custom ones. First, you have an existing model that you can see in the flesh. And you don't have as many decisions to make because many have already been made by the builder. At the same time, you don't have as many choices because the builder has already done the choosing. Your main involvement is in deciding which house you want in which tract. But don't make that decision from a position of ignorance. Some tract builders do a much better job than others of thinking about the home user. When you know what to look for you'll quickly figure out which builders and which plans to avoid.

In reality, it's not often that your only decision is which floor plan. First, there's the lot—you usually have a choice of several on which the house can be built. Then, there are the options, frequently many of them, that the builder is offering. And, finally, there may be things you can have done that are not in the standard option package, things that will make the house more cost-effective and a better place to live.

A word of caution here. I've seen many tracts where modern marketing techniques have taken over and, to get the base price down, builders have left out as much as they can and still have a house. They make it up with the options which everyone wants. (Would you, for example, want a house without an electric garage door opener? But many tracts don't have any!) Be extremely cautious when comparing tract models, it's often the cost of the options that will control what you spend.

Tract houses are known in the trade as "production" houses because they are built on more of a production-line basis than are custom houses. The builder, why may also be the subdivider, plans how many houses of each of several plans are going to be built and then buys materials in larger quantities. He is also in a position to offer subcontractors continuing work which means he can negotiate lower labor costs. But the more he does this, the more rigid he has to be about not making changes.

In every case, it'll pay you to be knowledgeable, not ignorant, of good and bad design and construction.

Custom Houses

For a custom house, where do you start? There are so many things that seem to need doing first: finding a lot, a house design, financing, getting a builder—the list seems endless and all are big decisions. At this point, you might be tempted to back off and buy that "ho-hum" tract house in the local subdivision.

You shouldn't. Getting a custom house is well within just about anyone's ability. But there is work involved. You'll have to figure out just what you want, not as a nebulous dream, but as something that can be put onto paper and then built.

Among the reasons people buy custom houses:

1. They want a truly one-of-a-kind house.
2. They've found a special lot, often with a great view.
3. They're in an area where there are few tracts and these are all low-priced starter homes. So the choice is either a spec house or

Nobody said it wasn't work.

a custom house and, since most builders make both, buyers can get what they want if they have it custom built.

4. They want to do part of the work themselves and can work out a deal with a builder who will let them do this. Or they can act as their own general contractor and do as much themselves as they want.

5. They're willing to spend the time to shop for appliances, fixtures, cabinets, countertops, etc., and save significant money in the process.

Besides the usual process of hiring an architect or building designer and a builder, there are several alternative ways to get your custom house built that may appeal to you.

Do it yourself—One of life's greatest satisfactions can come from starting from scratch and ending up with a home that's the result of your ideas and decisions. You can even be your own general contractor! This may seem ridiculous to you right now, but it's not that difficult with the help of companies formed specifically to help you make decisions that general contractors make and to help you save part of the money that a contractor needs for overhead and profit.

Kit homes—Depending on where you live, companies will sell you, delivered on-site, all the materials you need for your house along with detailed instructions about how to erect it. They'll either sell you a pre-designed house (with its shortcomings) or will engineer a kit to your specifications (based on their standard models). You are responsible for making the arrangements to have the house built.

Modular homes—If you live where they are available, modular homes are another way to get the custom dream you want. Modular homes are "stick built" houses that are made in a factory and hauled to your lot. They are **not** glorified mobile or manufactured homes. They can be fully customized including design and materials. They have several advantages over the traditional way of building:

• They are less expensive,

• Workmanship is likely to be more consistent, and

• They can be built and erected in days, not months.

But *Caveat Emptor*—neither kit nor modular suppliers are into user-friendly designs. You need to know what you want and then make sure you get it.

Which Is It?

Be careful if you're planning to get a custom home. There are at least three different ways that builders and their agents throw out the word "custom" as an enticement for the unwary when, in fact, what they are offering is not custom at all. Here are three examples:

- A kit home in which the design is fixed and the supplier will not change it. (Not all kit homes are this way.)

- A home built on your lot by a builder from a plan that you select from those that he offers. No significant changes are allowed. (All builders that work this way are adamant about not making changes. Be sure you understand the situation before making a commitment.)

- A subdivision which says "custom" homes but that's the only thing about them that is custom. They justify the use of the word by saying they include custom *features* —like tile rather than laminate countertops. They may be more upscale but they're still production houses with no significant changes allowed.

The Choice Is Yours

Whether you decide to go with a custom house, tract house, or spec house, some things are clear:

- Making sure your home is user-friendly is up to you. It is not a usual part of the way that home designers, real estate agents, or builders do business.

- You can get more value in your home with little more cost when you know what makes a house more livable.

- It'll be worth the effort you put in it both while you live there and later when you're ready to move on to an even better dream.

More Help

Feedback from readers indicates that checklists are invaluable tools in keeping track of and applying the information in the book. The original design checklist has been updated for use with the revised edition. It is available from Home User Press. Ordering information is in the back of the book.

Chapter 2

The Home-Building Community

When you bought your first home, it probably involved a real estate agent who acted for you in the process—or so you thought. That may not have been the case but at least you trusted him/her and, if it wasn't all to your advantage, you never knew.

Our lack of knowledge about the people and organizations who are involved in the biggest financial operation most of us will ever encounter is abysmal. This chapter won't keep you from making mistakes. But it will give you some background for your dealings with the building community.

Not all of it may apply to your circumstance this time, but you'll be selling and buying again. To have an understanding of what you're up dealing with is always helpful. Take the time to read through the chapter. If nothing else, you may find the answers to some questions you've often wondered about but never asked.

Four different groups are discussed here: real estate agents, designers, builders, and code enforcement officers.

Trade Associations

Trade associations (groups of people plying various trades) have existed since the days of the medieval guilds and are most often beneficial. By focussing efforts in areas of their trades that individuals could not or would not do, they can bring about changes that benefit everyone.

But do not be mislead to think that they are there *primarily* for you, the customer. First, to do anything, they must keep themselves in existence and to do this they must serve their members. It is for their *own* benefit and protection that a bunch of competitors get together to form a trade association.

This shows up in several ways, one of the most significant being the political pressures they bring to bear at national, state, and local levels. It is to the benefit of all of these organizations that people keep buying new homes. So, when it is suggested that something be done to control urban sprawl, you can bet the political action committees will spring into action to let nothing, from high interest rates to taxes to growth boundaries, stand in the way of building more homes. For the industry it's a natural thing to do—it's called self-preservation.

But there is a darker side, too. When it should be a matter of self-policing, too often trade organizations will stand up for someone they should be riding out on a rail. And this, in turn, can lead to using the strength of an organization to promote and defend laws that are narrowly self serving.

Personal note: After the above was written, I came across a situation where a group of builders had their memberships in their local builders association (and hence in the NAHB) revoked because of unethical practices. There **are** standards that can be enforced.

There are three trade groups that, even if you don't come into direct contact with them, will have an impact on you and your new home: the designer, the builder and the sales agent. Their organizations are discussed here.

Real Estate People

Real estate brokers and agents are licensed by the state. They are obligated to obey state laws. Brokers assume the ultimate responsibility for real estate deals—agents must get the broker's approval before a deal is finalized. It's a way to double check the legalities involved. Agents work with sellers and/or buyers in arranging sales or trades of real property, including houses.

A broker is bonded and must have several years experience as an agent before being eligible for a license. The owner of a real estate firm is usually a broker; large firms will have several brokers. The relationship between the firm and the agents is not that of employer and employees; rather, while agents are affiliated with their firms, they are independent businesses in many ways. Competition is stiff. Agents in the same office compete with each other (and all other agents in the community) to get listings and to represent buyers.

Most real estate brokers and real estate agents belong to a local Board of Realtors. In each state there is a state association of realtors and nationally there is the National Association of Realtors (NAR). This politically powerful organization with over 800,000 members is the world's largest trade association.

The national association has a published code of ethics for its members. This code encompasses the relationships and interactions between realtors and between realtors and their clients. If you get a chance, take a look at it. It puts pretty stiff constraints on what agents and brokers should do.

You'll see the initials GRI after some realtors' names. Graduate, Realtor Institute is awarded to a realtor after the completion of three one-week courses offered by the state association. This program is for brokers and associates who are involved with single-family homes.

A Certified Residential Specialist (CRS) indicates that an agent has taken extra studies in this particular area. (Aspects of home design are *not* a part of these studies.)

The local realtors' association usually sponsors a multiple listing system (MLS) in which all properties being sold by the members are put into a computer file that is accessible to the members who subscribe to the service. It gives the local realtors the ability to sort through the listings and come up with those that closely meet the needs of a client. You'll probably, in the not too distant future, find the MLS listings available on the Internet.

"Realtor®," "Realtors®" and "Realtor-associate®" are all registered trademarks of the National Association of Realtors. Only members of that organization are entitled to use these trademarks. Others should be referred to as real estate brokers or real estate agents. Note that, like many other trademarks, the terms are often used in a generic sense, that is, "realtor" instead of real estate agent, etc.

In the process of selling and buying a home, different agents usually represent the seller and buyer. Depending on the laws of the individual states, the agent that the buyer thinks is representing him may, in fact, be a sub-agent of the seller's agent. This puts the buyer at a significant disadvantage in the buying process. Some states have changed their laws so that a buyer's agent must truly represent the buyer and the laws are being changed in others.

But there is one aspect of it that you should never forget. As long as the agent representing you, the buyer, is on commission, it is to his/her immediate advantage to make the selling price as high as possible—not to your advantage, for sure. (The other side of this is that an agent's long term success will depend on referrals from previous clients and to this end it is advantageous to do as well by a buyer as possible.)

One of the problems real estate agents have in representing buyers is that the buyer is under no obligation to buy through an agent even if the agent has spent many hours showing properties to them. To get around this, you may be asked to sign a contract with an agent in which you agree to pay him/her a fee whether you buy through him/her or not. Don't sign it until you are confident that the agent is competent, not only with regards to the legal and financial aspects of the transaction but also that he/she knows a good property from a bad one. Most do not. A good agent will be well worth the money. A poor one is a waste of time as well as money. Shop around and get references. And, even then, be sure you have a reasonable escape if things don't work out as you had expected.

Recognizing a market need, another type of real estate company has come into existence—the exclusive buyer's agency. These companies do not take listings; they only work with buyers. Their agents are paid a fixed fee—not a percentage of the selling price of the house. They are not salespeople but rather consultants and advisers to buyers. The exclusive agency concept is more strongly developed in some parts of the country than others. The national organization, the National Association of Exclusive Buyer Agents (NAEBA) with offices in Evergreen, CO, (800) 986-2322, can send you a list of members. The Web address is http://www.naeba.org.

But do be careful here, too:

- Most NAEBA members are no more qualified to warn you about defects in the way a house is designed or built than their non-NAEBA peers.
- Be sure the agent's fees are independent of the selling price of the house so there is no incentive to keep the price high. Responsible buyer's agents expect this kind of an arrangement.

Home Builders

Most builders belong to a local or regional builder's association that is affiliated with state and national associations. These associations have the words "home builders" or "building industries" in their names. The National Association of Home Builders (NAHB), is the trade association for the industry at the national level.

The associations, at all levels, represent members before various governmental agencies. They sponsor and work for legislation benefiting the building community. Membership comes from all areas of the building trades. It includes remodelers and small, medium, and large volume builders. Associate members include subcontractors, materials dealers, house designers, lenders, and realtors.

Membership in the local association is seen as giving credibility to the member. The building community recognizes NAHB membership as a sign of professionalism.

At the local level, the home builders' associations provide group insurance programs, sponsor seminars, and hold periodic meetings and social events. These activities are for the purpose of education, business contacts, and general assistance to the members.

On the national level, the NAHB tests and approves new building materials. It has a number of services available to its members that are intended to provide help in areas of home building. It lobbies at all levels of national government to "ensure housing remains a top priority during the formation of national policy." Individual members are asked to bring the NAHB message directly to their representatives in Congress. The NAHB has the third largest trade association Political Action Committee (BUILD-PAC) in the country, raising over $2,000,000 every four years.

It publishes monthly magazines along with other data available to the members upon request. The NAHB sponsors educational seminars and the country's largest trade show.

Builders aren't required to belong to the local association. A few don't.

Designers

There are two organized groups whose members design houses, the American Institute of Architects (AIA) and the American Institute of Building Designers (AIBD). And there are many designers who don't belong to either of these.

AIA

AIA members are registered architects in their respective states. To be registered, an architect must have graduated from an approved university-level school and passed a stringent state exam. Both by their education and general interest, architects are generally capable of designing buildings of much greater complexity and size than single-family residences. Most architectural firms do residential designs but only a few specialize in them.

Some architects do not work for the larger firms but have chosen to be on their own and to specialize in residences. You may find them more attuned to the "good taste" aspects of design as well as being more thorough than those who do not have an achitect's education.

AIBD

AIBD members may be registered architects (or even AIA members), graduate architects who aren't registered, experienced designers, or neophytes. There is no state registration required to do what AIBD members do. The only indication of their competency is a certificate issued by the AIBD when members meet AIBD requirements for experience and then pass a set of AIBD tests.

Which?

Generally, if you hire an architect, you can expect to pay more to have the work done but it is less likely to be a commonplace design. Architects will put more detail into their plans, leaving fewer decisions up to the builder and his subcontractors. Architects will usually, for a fee, keep an eye on the house during construction. This may be helpful if there are unusual design features. Depending on what your new house is to be like, you may be better off with an architect or you may find a less expensive building designer is just fine.

Two things to be careful of:

1) There are good and bad, both architects and designers, and

2) Neither is likely to be sensitive to user-friendliness in their designs. That's where and why you should plan on being involved in the process.

Code Enforcement

You may not have thought of them as a part of the building community but the officers/inspectors who enforce building codes may have a significant impact on what your new home is like. As this book is being written there are large upheavals in the whole code

area. The most significant is that a standardized residential building code (CABO, *see* Appendix B) is being pushed for adoption in
all states.

Some states, notably those on the west coast, already have statewide codes. Most states, however, leave it up to local jurisdictions—
counties, townships, boroughs, cities, and towns—to adopt codes
to control what's being built. Some places have no codes at all
and in many locations where there are codes there is little, if any,
enforcement. Builders can then build what they want—which is
too often what they can get away with.

Even after a more general adoption of CABO, the matter of enforcement will be left to local jurisdictions so that the number and
competency of building inspectors will still be a matter of the whims
and budget realities of local politics.

There is little you can do about all this. But it sure behooves someone who is making the investment in a new house to know what
the rules are and to have some idea of where you can turn when
you feel that your builder may be trying to cut corners. If you live
where code enforcement is scant or nonexistent you may want to
consider having a clause in your agreement with the builder that
you can hire a competent inspector to check that he is following the
CABO code—or whatever may be in effect in your area.

Chapter 3

Lots

Wild fire, wind, flooding, tornadoes, earthquake, hurricanes, winter ice, or summer sun—which is of more concern where you're thinking of buying? Or is it the CC&Rs, the building codes, or the price? Every one of these should be taken into account when you select the lot for your new home.

But, above all, location—length of commute to work, suitability of schools, neighborhood, distance to shopping, and taxes. Some lots are special because of the view they offer. Other locations may be important for personal reasons.

In this chapter several additional aspects of lots are considered, aspects that are easy to overlook in the enthusiasm of getting a new home. The same considerations apply whether you're getting the lot because you're going to build on it or if it's a lot in a tract where your new house will be. If it's a spec house, you may well change your mind when all of the lot considerations are taken into account.

The difference between a purely spec house and a custom house can become blurred when dealing with lots. If you find an existing house and want one just like it, you work directly with the builder. With the lot you, may buy it on your own or it may be one the builder already has.

Subdivision Constraints

Whenever you buy a lot in a subdivision, the CC&Rs (Covenants, Conditions, and Restrictions) can be crucial. (These are discussed in detail in Appendix A.) They restrict what you can do on a lot, even what kind of house can be built on it. You'll have to sign a document acknowledging that you agree to abide by the CC&Rs

before you get title to your lot. DON'T WAIT UNTIL THEN TO READ THEM.

When you're looking at a lot, the sales agent may not tell you that there are any CC&Rs. You should ask and get a copy if the lot interests you.

At other times agents may make a big thing of telling you how great the CC&Rs are, how they really add value to the property. What they may not tell you is that no CC&Rs are worth the paper they're written on if they're not enforced. (Many CC&Rs are for the convenience and protection of the subdivider and, as such, are worthless to the homeowner.)

Before you buy your property, read Appendix A and then read the CC&Rs. There might be some constraints, or lack of them, that would make the property unacceptable to you. As a final step, check with other homeowners in the same subdivision and see just how effectively the CC&Rs are enforced.

The House and the Lot

Visualize how the house will fit on the lot. Usually there are local constraints, either CC&Rs or ordinances, on the amount of front, back and side setbacks required. Consider the driveway and the front walk. If you want a garden or an RV pad between your house and the adjoining one, is there room?

If there are, or will be, houses on adjacent lots, check how much privacy you'll have. Wooden fences, common in the western states, might be all you'll need. When the adjacent house has two stories or when it's above yours on a hillside, neighbors may have a great view of your back or side yard—and into upstairs bedrooms.

In coastal areas there may be persistent on-shore winds, especially in the afternoons. These are often strong enough to be downright annoying, particularly when you have to fight them for several months every summer.

Lots with great views can have a similar problem. They often sit on a bluff overlooking a scenic river valley. The problem is that such views are often just where winds blow almost constantly. Strong winds blowing against windows cause noise and, when the winds are continuous, the noise can become depressing. (These considerations can be a constraint on what kind of siding you are going to have on the house because vinyl siding can rattle in the wind.)

At the other end of the spectrum are lots alongside an easy meandering creek where the most threatening things are the frogs that croak all summer. Then comes the once-in-a-lifetime flood. If you're lucky you'll have flood insurance. If you're not, cleanup can be costly. The best thing, of course, is to not build there in the first place.

Owning a house in a low-lying area is always risky.

If you've already decided on a floor plan, check the orientation of the house relative to the sun and the street. The street usually dictates how the house will sit on the lot. One school of thought is that the kitchen should face east. North-facing kitchens lack sunshine all of the time, while south- and west-facing kitchens get the summer sun when you may not want it. Sometimes reversing the plan will make the arrangement acceptable; other times you may decide that the lot and the house you want just aren't compatible.

If you pick a lot for its view, find out what protection, like CC&Rs or local ordinances, you have to prevent someone from building in front of you and blocking your view.

If your prospective lot is close to farming areas, take a drive around to see just what kinds of farms may be there. Dairy farms generate odors that even gentle breezes can waft for a mile or so. Similarly mushroom factories, which use manure to grow their products, can be a source of unwanted smells.

Hillside Lots

Hillside lots are attractive because they often include a view. You can expect to pay a premium for these lots in two ways: 1) the price will be higher than comparably sized lots and 2) you'll have to move more dirt and pour more concrete than normal. If the lot is remote, there are other considerations that we'll discuss later. A builder can give you some idea about the extra cost.

Hillside lots are a real danger in areas where wild fires may occur because fire races up hills and can be uncontrollable. A more subtle problem is that fires may not destroy your home but, because they destroy the vegetation, they increase the likelihood of mud slides the following rainy season.

Some areas are known for the number of houses that have slid down the hills over the years. If you want to build there, be sure to have an engineering study done to find out if it's possible to anchor the house and what the costs will be for doing it.

When a house is on a hillside lot and the garage is cut back into the hillside, you may be giving up a side garage door. You should factor this into your planning.

Ground Water

How can you tell if ground water is present? If the lot hasn't been graded clear of vegetation, be suspicious when you see a heavy thicket of greenery. Ask to see any engineering reports made on the lot or the subdivision. Talk to neighboring property owners to see if they've had water problems during the rainy season.

The presence of ground water will often mean extra expenses for foundation and drainage. And, if not taken care of properly, it can be a source of problems later on.

Grading and Drainage

You'll need adequate drainage from the lot. If the lot is below street level one of two things are needed:

- A sump pump to get the water back up to the storm drain in the street. Sump pumps are noisy and cause maintenance headaches when they're in the crawl space.
- An easement so that a drain line can be run across an adjacent downhill lot to a street below. Check to see if the easement exists.

Also, check what happens to water from an adjacent lot located above yours. If it runs on to your lot, ask what's being done about it. If you're not satisfied, find another lot—or you may be faced with a lake in your yard when the rains come.

Another concern you should have with the lot is the steepness of the driveway into the garage. Steep driveways have two problems: 1) if there's an abrupt change in slope, part of your car will drag and 2) in icy or snowy weather you may not be able to use your garage, particularly if it's facing north.

Above-Ground Utilities

When a plot of land is being subdivided sometimes there will be a set of power poles running along one or more sides of it. Lots along those sides will then have the power and telephone lines in front of their houses. We lived this way for many years but not now. If you make an offer on such a lot, take into account that its future value will be lowered because of the power poles.

Utility Boxes

Utility boxes are often forgotten when people look at a piece of property. In new developments with utility wiring run underground, the power company must put their transformers someplace. In some areas these may also be underground but, if they're not, they're an eyesore, especially if the box housing the transformer is in front of your house. Above-ground boxes for TV distribution

Utility boxes don't improve the appearance of a house.

amplifiers and telephone junction boxes, while not as big as power transformers, will also look better down the street.

You don't have much to say about where utility boxes will be located, but at least you can see where they are when it comes time to select a lot. Likewise, groupings of mail boxes furnished by the Postal Service aren't things of beauty and would be better placed away from your lot. When it comes time to sell, prospective buyers will see things like this.

Far-sighted land developers can do something to reduce the visual impact of these functional things. In some tracts subdividers build housings around utilitarian-looking mail boxes. These cosmetic housings improve the appearance of the boxes and have an impact on the perceived value of your property.

Traffic Noise

Traffic and traffic noise will affect the selling value of the property.

We moved into our house in northern California when the nearby freeways had little traffic. When we put the house on the market 15 years later, we had more than one prospective buyer pull up in front of the house, stop, listen, and get back into the car and drive off.

Take the time to find out what future traffic patterns are planned around any lot you consider. Stay as far away from freeways as you can and at least a block away from what could become a major arterial.

A couple we know chose a lot in a large subdivision when the builder was just starting construction. It was close to the golf course clubhouse, making it convenient for golf, dinner at the club, and the other social amenities they desired. The lot backed up onto a street that had no traffic; in fact, it dead-ended just a block away—a quiet, ideal place to spend their retirement years.

Over the next several years they found out what the sales agent hadn't told them. Their quiet dead-end street was no longer quiet or dead-end. As the subdivision grew, so did the street, both in length and in volume of traffic. It is, in fact, a very busy arterial. They have decided to put up with the noise rather than go through the stress of moving again. But the reminder is always there—you can't be too careful when choosing the location for your home.

Outside noise can come from sources other than traffic. Railroads, even those several blocks away, can generate noise you'd rather not have. Check how close the lot is to an airport, commercial or military. Even though planes may not be flying overhead when you visit the lot, changes in weather can cause changes in air traffic patterns. It's not what you need in your nice quiet neighborhood.

Corner Lots

There are pros and cons about corner lots. You have one less abutting neighbor but you also have two sides of the house facing a street, making for more maintenance of lawns and shrubs; at the same time, your backyard becomes visible from the street.

These lots are favorites for people who cut corners, either on foot or on bicycles. It may take a fence to stop them (if allowed by local ordinance and the CC&Rs) and a fence may not be aesthetically pleasing to you.

On the positive side, a corner lot gives you two ways to access a street. For example, your garage can be facing the street in front of the house while the driveway to an RV pad can be on the side street.

If you buy a corner lot and the front of the house faces the same street as the garage, be sure the garage side of the house is away from the corner. You may have to reverse the floor plan to do this. There are several reasons:
- It's safer to back the car out of the driveway.
- The side of the house away from the garage is often more attractive from the street.
- The garage side is where you're more likely to have an RV pad and garbage cans—items you'd rather not have on the street side of the house.

And corner lots are but one example of where headlights from turning cars can be a problem.

Lots and Headlights

While driving at night in a residential neighborhood watch where your headlights hit, particularly on high beam. When they hit the front of a house with sidelights around the door, there will be a flash of light in the entry. When they hit a bedroom window, special attention to window coverings will be needed if sleepers are not to be bothered.

Curving streets and intersections have the potential for this problem. When checking out a lot, see if cars coming from either direction will be directly facing the side or front of the house. If so, you'll need to be careful of window placement in bedrooms and entries.

While upstairs bedrooms won't generally get the full brunt of headlights, they will get some of it. And, if the street is on a grade, even they may not be spared.

There are three types of corner lots to watch for:

- In a common four-way intersection, all the corner lots will have headlights swing across them as cars turn. Even houses on lots next to those on corners may be affected.

- When the lot faces the end of an intersecting street, you'll have to be careful of the house plan you select because headlights of cars coming up the street will hit the front of your house. And you may have second thoughts, too, about the possibility of a driver who is inattentive and forgets to turn. In general, these are not the most desirable lots on the block.

- About as bad is an L where both streets end. Houses on corner lots will face down one of the streets, with the same considerations as the previous situation.

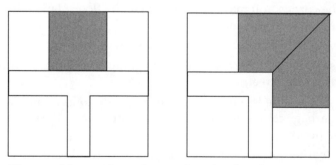

Give special attention to house plans for these lots.

A cul-de-sac is attractive for young families because the closed end can be a relatively safe place for the kids to play. But, for lots at the end, headlights will swing across the fronts of the houses as cars loop around the end of the cul-de-sac. You probably won't want bedrooms there.

Zero Lot Lines

Some subdivisions use zero lot lines, meaning one wall of each house is located on a property line. In some cases, two houses share a common wall, so that from the street the two houses look like a duplex. In other cases, each house is separate, with all houses in the block placed on the same side of their lots.

If you consider a house on a zero lot line lot, be careful to understand all that it entails. There will be some definite restrictions on what you can do including a requirement that you use the subdivider as the builder of your new home.

A monetary caution about zero-lot-line tracts. They are, of neces-sity, associated with a homeowners' association which is responsible for street-side landscape maintenance and fire insurance. The association dues can be sizable.

Country Lots

When the lot you're considering is outside a metropolitan area, you need to remember water, sewer, gas, electricity, telephone, and roads. If you have to drill a well, you need a back-up plan in case the water you get is not potable. If you require a septic tank and leach field, you need to know the soil is suitable. The seller may be able to guarantee these things; otherwise make any offer on the lot subject to your approval of test results.

But before you do that, find out what the utility companies will charge to get power, gas, and telephone to the lot. Also, find out what the costs will be to have a road extended to it.

Owners of houses in the country, or even in the city next to an undeveloped area, need to be concerned about the danger of wildfires. In Oakland, California in 1991 fires destroyed many hillside homes and, as a result, a number of jurisdictions in the state enacted special local building codes. These require that homes built where there is a danger of wildfires take extra precautions to prevent the house from burning. If you are going to live where this may be a concern, be sure to check what it may mean in extra building costs. It can, for example, be prohibitively expensive to run a water line for a fire hydrant to your lot. You'll need a non-flammable roof—and that, too, will add to the cost.

Easements

Find out if there are any easements on the lot and, if so, what they mean to you. Sometimes it means you have to give up any use of part of your property; other times it means you can't use it for some purposes. Find out before you buy.

Don't be like one couple I talked to who looked at a lot with a real estate agent. The lot and the neighborhood suited them fine. Sure there was this ditch across the back of the lot, but they'd get that filled in when they did their landscaping. It was later that they discovered the ditch, which is six feet wide and three feet deep, is covered by an easement. They can't fill it in because it's for rain water that flows all winter. They're now looking at an expensive drain if they want to use the rest of their lot.

Special Assessments

If you're buying in a state where they may exist, be sure to check for "special assessment" or "local improvement" districts. For example, in California local jurisdictions can set up special assessment districts so that new home buyers pay for the streets, sewers, and other utilities to service new subdivisions. This seems fair enough. But when buyers first learn about it when they get their tax bill and find a large payment due on the assessment—that *isn't* fair!

And then there were the folks in Vancouver, Washington who found out in 1997, a year after they moved into their new homes, that there was a $4000 local improvement district assessment on their lots to pay for new roads.

It is a shock to new owners to find they owe a large, totally unexpected, special assessment. Not all sales agents or builders are quick to volunteer the information—it could make you decide to look elsewhere. Ask!

Chapter 4

Home Designs

I keep remembering an article in the Sacramento Bee about the study I did on design flaws. On the same page as the article was an ad promoting a tract house with a floor plan that had two of the flaws discussed in the article! (This isn't unusual because flawed floor plans abound.)

Our need to assert our individualities leads to a tremendous diversity in houses and house plans. No two are alike, even though they may be the same model in a tract. (Essentially identical houses become different when the floor plan is reversed or the exterior is changed.)

You may wonder where all these designs come from and why they're not all user-friendly. Understanding the answers to these questions will help you know what you're dealing with when looking at houses and designs.

Builders work from designs; they don't "wing" it—if for no other reason than that they can't get a building permit without plans. The people who make plans, architects or designers, are not alike—neither in what they do nor in their objectives.

Designers

There are several different design business areas and some designers work in more than one of them. The discussion here is meant to help you understand what you're looking at when you see a fully custom plan, a plan in a newspaper or magazine, or one in a catalog you pick up in the supermarket or bookstore. Note that home designs can be drawn by anyone, qualified or not. He/she may be a registered architect, an experienced designer or a beginner who works at it evenings and weekends. Or it may be someone who doesn't know what he/she is doing and does it anyway.

The Mass Producers

If you haven't done it already, take a look at any national home magazine and you'll find full page ads for catalogs of home plans. Or you'll find pretty pictures of houses with floor plans—along with offers to sell you the whole set of plans. These marketers gear their products to what they can sell—to what they think individual people are going to buy. For this market, repeat business is not a major consideration (after all, once you've built your home you're no longer interested in plans) so what goes into the plans is what the publishers think will sell right now with little concern about your future. There is no incentive to make plans other than what will appeal to most people—or to a specific group of people like those who read magazines on country living. And there is certainly no reason for plan producers to be concerned about user-friendly designs.

The "Better" Designers

Then there are designers who cater to builders rather than the general public and make plans for builders to buy. The plans they make are called "stock plans." Their buyers are the spec and small tract builders. They, too, sell catalogs but not like the ones you see in grocery stores or advertised in national home magazines. These catalogs are more expensive, typically selling for several hundred dollars apiece. But, like the mass plan producers, their whole objective—and the way they make their livings—is to sell plans. Repeat business is important and more attention is paid to the quality of the plans—at least in what will appeal to builders. But, since user-oriented designs are not, in general, important to builders, neither are they to these designers.

Designers for Tracts

Large tract builders have their own design staffs or retain a large design company. The objective is not to sell the plans themselves but to make designs that will satisfy the market where the tract is to be built and make it possible to build the houses at the most competitive price. These designers get their guidance from detailed market studies of the area around the tract and of the population of expected buyers. Their designs are an excellent reflection of what people want. It's a whole different business than the two earlier categories of designers.

Not all tract builders, however, have the resources to have a full-time staff of designers. Smaller, local builders may buy their designs from catalogs but are more likely to take someone else's design and make changes they think will make the house sell better and be cheaper to build. As noted below, be especially careful with these builders and their homes because they're frequently not done well.

The Custom Designer (and Redesigner)

The custom designer provides a service. He/she works with a client to find out what the customer wants and what can be afforded. The object is to satisfy the customer. Advice and suggestions may be proffered, but the end product is the result of customer inputs. In general, the designs are user-friendly only if this is important to the buyer.

The Others

There are people who do plans as after-hours projects. These include builders and "draftsmen" who aren't qualified or can't make a living as designers.

A word about builders who do their own plans—and this includes small tract builders. Without a doubt the worst designs I have seen—and I've looked at over 2000 of them in the last five years—are ones that have been done by builders who seem even more removed from the end user than professional designers. The diagram, for example, was from a house whose plans were made by the builder.

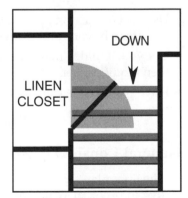

A door that swings over the stairs. Incredible!

Designs

Floor plans are scale diagrams of the layout of the rooms, halls, etc., in a house. There will be separate drawings for each floor.

House design packages or house plans include floor plans along with the many other drawings needed by a builder to build the house. (But not all packages are equal. Some include considerable detail to ensure that the builder makes what the designer had in mind. Others leave a great deal to the builder's discretion.)

There are several distinct ways you can go about getting the design for your new home. The simplest is to find a house you like and have the builder make another one just like it. This is what happens in a tract or when you find a spec house that you'd like to see on your own lot.

But you'll probably want the builder to make a change here and there and some tract builders may encourage this by offering an extensive package of pre-planned options. If you take this route to your new home, don't be discouraged by not finding exactly what you want. Some smaller tract builders are more than willing to consider changes other than their standard offerings.

We did just that in our southern California house. We liked the floor plan and the way the model was finished but decided we'd like some changes besides those the sales agent offered. The builder was more than cooperative. Sure it cost us more money, but he was happy and so were we.

If you decide to have your new home custom built, you will need a set of plans for the builder to work from. Your options are to:

• Use a set of existing plans from a magazine or catalog. Whatever changes you want made are done by you and/or the builder. You don't use a designer.

• Use a set of existing plans and have them modified by a designer. Be sure to be careful of copyrights, it's easy to infringe on them.

• Get your builder first and choose from his available plans. He then modifies them as you wish.

• Have plans made from scratch by an architect or designer.

Floor Plans

General Considerations

The floor plan should make the house feel comfortable and easy to live in. Look at which areas will get the most use and where people will walk when getting from one area to another.

Here are some things to remember. As you become aware of them, you'll no doubt find others you'd rather avoid.

- Is the traffic path through the kitchen where culinary efforts are under way? Is there a lot of traffic across the family room in front of the TV? It takes only a few minutes to decide if a floor plan makes sense in terms of how the house will be used.

- Consider the room arrangement and noise. Family rooms next to bedrooms can be a problem unless the wall between is sound insulated.

- Is there an entry closet? Surprisingly, some designs don't include them. The same thing goes for linen closets. These are often in hallways or, sometimes, in bathrooms. And, in some cases, there aren't any at all!

- Is the garage close to the kitchen? When it is, it makes for shorter trips from the car to the living area of the house, particularly with groceries and trash. In some house designs the garage can do a good job of shielding part of the house from midday or afternoon sun or winter storms if it's on the right side of the house.

- If you're planning a two-story house, be sure to include a half-bath downstairs.

- Bedrooms over the garage are virtually impossible to insulate against the noise of the opening and closing of the garage doors. Think twice when you see this arrangement.

- Is there a door out the back of the house? Surprisingly, some plans forget this. In other cases, the way out to the back is through the garage but the garage lacks a side door!

- One consideration in choosing your floor plan and house design is whether it lends itself to future expansion. This may be important to you or, at a future date, to someone else when they're looking for a place to buy.

- Some people object to being able to stand at the front door and look through the house to the kitchen. If this bothers you, watch for it. (Split level entries are often this way.) Or, since it could reflect on the resale value of the house, just don't have a house where this can happen.

- Two other potential problems with split levels are inadequate space at the front door (you should have room for both you and your guests to comfortably stand there) and stairs so narrow that it is virtually impossible to move furniture up and down. And watch where the guest closet is located. There is a half flight of stairs between the door and the closet. Don't have it any farther away than is absolutely necessary.

- In the same category as being able to see the kitchen from the entry are objections to arrangements where you can see into a bathroom from the living area of the house. Even if you don't care, a potential buyer someday might.

- Is there enough storage space? Because it isn't a showy thing, designers and builders often skimp in this area. Be sure there are places for brooms, vacuum cleaners, and things you may never need but want to save anyway—like the original boxes for electronic equipment.

  Don't forget a place—a basement is great if you have one—to put seasonal items from skis to snow blowers and lawn mowers. Attic storage can be useful but needs to be planned (*See* Chapter 12).

- The kitchen should be next to the formal dining room, if there is one. Separation of dining room and kitchen means many extra steps when serving food and cleaning up.

One Story or Two?

In deciding between a one- and two-story house, consider:
- Stairs.
- Space taken by stairs and outside walls.
- Heating and cooling.
- Lot requirements.
- Shading from roof overhangs.

Stairs—Older people, even those without physical disabilities, prefer not having stairs. Two-story plans with the master bedroom on the first floor are helpful but most often single-floor arrangements are the choice of senior citizens.

Space Taken by Stairs—A typical stairwell, including framing, is 4×13 feet which is over 100 square feet (50 on each floor) of space that isn't needed in a one-story house. Landings also take away from the useful floor area.

Overall, the single floor house will often be higher priced for the same amount of square footage. But don't assume that a two-story house is a better buy until you take out the floor space used by stairs, landings, and outside walls.

Outside Walls—The area on each floor taken up by outside walls is included in the square footage of the house. (Square footage is measured on the outside, not the inside, of a house.) With two floors this lost area occurs twice, once on each floor. In a typical 2400 square-foot house, the floor space taken by the walls will be about forty square feet more for a two story house than for a single story with the same number of square feet..

Heating and Cooling—As you know, upstairs rooms are noticeably warmer than downstairs areas. Most builders don't do anything about this even through there are ways to correct it. (See Chapter 5.) It will cost money and this should be taken into account when comparing one- and two-story houses.

Lot Requirements—For the same amount of useful floor area, the two-story house can take up less of the lot. If this is important, it may be the deciding factor in selecting your floor plan.

Shading—Roof overhangs can be used to shade windows on one-story buildings but not the lower walls and windows of two-story houses. This can be important in the desert and other hot climates.

Simplicity

In general, the simpler the outside of the house, the less expensive it will be. A rectangular box will be more economical to build than a house with notches and zigzags in the walls. Similarly, the more complex the roof line, the more costly it is. Breaks in the roof line, gables, and turrets add to the cost. Steeply pitched roofs will be more expensive than those of moderate pitch because of the greater difficulty (and danger) for the roofer. You have to decide how much money to allocate for the sake of appearance (and resale).

Architectural fashions come and go. In the past we've had baroque, rococo, and Victorian. Today, it's colonials, gables, glass, and white interiors—about as far from the plain rectangular box with its plain sash windows as you can get. In many designs each room seems to need its own gable, including the entry. Note how many new houses and designs have glass above glass across the front of the house. And the lack of continuity—they look as if the designer took pieces from wherever he/she could find them and stuck them together. Who knows which direction the pendulum will swing next, but history's lesson is that what's faddish today will be passè tomorrow. And that can be an expensive lesson at resale time.

The message: don't too get carried away with today's "in" thing. Have enough roof-line breaks to be interesting but don't overdo. Have enough glass to make the interior light and appealing but don't make it a fish bowl. Keep the interiors open for family togetherness but make sure there are private spaces too. And above all, keep it user-friendly.

How Many Square Feet?

Use the number of square feet shown on a floor plan only as a general guide. In many houses and plans there is useless space that still costs to build and heat—space that could often better be used elsewhere. The following addresses some of these areas. How they affect your ideas about a particular house design is an individual matter.

Conventional wisdom has it that a two-story house is less expensive per square foot than one story because of the smaller foundation and roof for the same number of square feet. This is true if you stop right there but it doesn't account for the non-living space taken up stairs and walls.

Stairs and walls aren't the only place where unused space is included in the overall square footage. Some master bedrooms have a large space in the center of the room whose sole purpose is to make the room look bigger. Even kitchens sometimes end up with space that isn't used. (This is unfortunate, because it will take more work in such a kitchen than would be needed with a better design.)

Other places to look for useless space are in the centers of walk-in closets and pantries. To be sure, you need to have a place to stand, but good design will minimize the amount of space required. Eliminating the walk-in pantry is one way to avoid the waste. Cabinet-type pantries with pull-out shelves make much better use of the space than does a walk-in unit, although they're not as pretentious.

Similiarly, eliminating the walk-in closet will also save space. While walk-in closets are popular today, wall closets have more useful space per square foot of floor than do walk-ins. (Eliminating walk-in pantries and closets also solves the problems of where to swing the doors.)

You'll often find unused space in hallways. Where a second-story bridge is needed across an open space between two upstairs bedroom areas you have an example of aesthetics first, followed by utility. (A basically different plan would obviate the need for the bridge. But it may still appeal to you—as it has to many others.)

Part II

The Systems

In this part the discussions are about the systems that involve the entire house rather than a single room or area. As with every part of the house, decisions must be made about the components in these systems. If you can make the decisions, you control what you get; otherwise it's up to the builder and the subs. And you'll get what they want—or don't care about.

The electrical, plumbing, and heating and air conditioning systems are described along with some aspects you need to know to be sure that you get what's best for you. Doors, windows, fireplaces and stoves, and security systems each have their own chapters.

The last chapter of Part II includes the important subjects of storage, noise insulation, floors, and ceilings.

A note about terminology: the almost universally-used interior wall covering is variously known as "sheet rock", "dry wall", "gypsum board", or "gyp board." The one and two family dwelling code (CABO) calls it "gypsum wallboard." We stay away from brand names and simply use "wallboard" throughout *BUILD IT RIGHT!*

Chapter 5

The Electrical System

This chapter deals with several distinctly different electrical sub-systems. With each you will need to be aware of good and not so good ways of equipping your new home.

As elsewhere, if your interest is in getting a tract or spec house, you should use the information in this chapter to grade different models you visit. It will also help in seeing things the builder might change for you.

For the custom home buyer, the details here are your guide to what is appropriate in your plans and specifications.

Entrance Capacity

The electrical system entrance capacity should be large enough for future expansion of electrical needs. In most houses a 200-ampere entrance is appropriate.

Switches and Outlets

Which type of wall switches will you want, the older flip switches or the more modern rocker ("decora") switches? They both work fine but rocker switches are perceived by many as easier to use and more in keeping with today's homes—a resale consideration. The rocker switches are a little more expensive but are an insignificant cost item in the overall. You'll want to consider dimmers for lights, too, particularly in living, dining, and family rooms.

Take a good look at the locations of outlets on walls where you'll be placing furniture. When they are behind headboards, sofas, or dressers they are a major problem for plugging in lamp cords. Or, if the electrician insists he has to put them there to meet code, then add one that will be easily accessible with the furniture in place. (Codes specify the maximum of spacing between outlets so that lamp cords aren't strewn all over the place. There is nothing that prevents putting outlets closer together.)

What color do you want the cover plates on switches, outlets, TV connectors, and telephone jacks? These are among the few things that can be changed later without tearing the house apart but it'll be one less thing to worry about if it's done right the first time.

Switch Location

Locations of switches are even more important than of outlets. Electricians sometimes have their own ideas and sometimes they don't have any choice. When framers put studs so close together that there is no room for an electrical box, the electrician has to do the best he can. Good framers look at the house plans to see where switches go and make sure they don't put lumber there.

In my house, for instance, the horizontal distance from the room entrance to the light switch varies throughout the house, more or less randomly, between 4 and 17 inches. It's a real pain trying to find a light switch in the dark.

Check where switches and electrical controls are located. You should be able to have the switches and wall-mounted controls for the various lights, fans, and gas fireplaces put where you want— electricians get paid by the number of switches, not by where they're located. As a guide, switches should be mounted in the room with the light, outlet, or appliance they control and the center of the switch should be within 6 inches of the entry to the room or hall. (Or at least they should be consistent throughout the house.)

One custom builder I know waits until the house has been framed then walks through the house with the electrical subcontractor and they note where switches and outlests are to be placed. If there is framing where there should be a light switch, that's too bad. The switch goes someplace else.

If the builder knows what you want and the framers leave no room for the box, it'll be up to the builder to get it fixed. (Actually, he'll be likely to be watching things a little more closely and won't let it happen in the first place.)

Have you ever gone into a bathroom and turned on the exhaust fan when you reached for the light switch? When they are mounted side by side this is easy to do. In some areas of the Pacific Northwest you'll often find fan switches located a foot above the light switch. If this doesn't seem too radical to you, give it a thought. While different to look at, it *is* more user-friendly.

This same approach is taken with ceiling fans: the light controls are mounted at normal switch height and the fan controls a foot higher.

A t first I was completely turned off by the appearance of these switches but, having lived with them for five years, I've come to appreciate it when you know that the light is going on and not the fan.

For gas fireplaces, the on-off switch should be close to the fireplace. It is not a good idea to put the fireplace switch together with the switches that control the lights because it's too easy to turn the fireplace on accidentally and leave it burning. If there is a separate switch or speed control for the fireplace fan, it should be in the same mounting box as the on-off switch—or mount it a foot above.

There is often a trade-off between aesthetics and functionality that is not clearly in favor of functionality. To you, the arrangement of the switches for a gas fireplace may not be as important as how they look. The photograph shows an example. While it is easier to remember which switch does what, who wants a wall that looks as if the switch locations were selected by throwing darts?

Decide how high off the floor you want the switches and outlets. 45 to 48 inches is normal for switches and controls and 12 inches for outlets. Again, they should be the same throughout the house.

Don't forget light switches in closets and pantries. These items are easily overlooked—before the

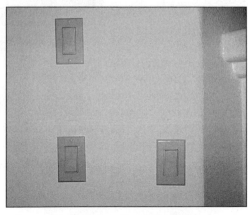

When a builder just lets it happen.

house is built. If you catch it in time you can have these added to an existing design. But be sure they're in a user-friendly place.

- In one house I visited, the switch is on the wall on the outside of the walk-in pantry but it's on the hinged side of the door that swings into the kitchen. To use the pantry, it is necessary to walk around the door after turning on the light. Not user-friendly.

- In another example, the light switch is inside the pantry where it belongs—except you can't see it because it's under a shelf! Be alert to keep such annoying and unfriendly workmanship out of your house.

- Above all, don't forget the light in a walk-in pantry. In one tract I visited I couldn't find the switch for the pantry only to discover that there was no light! At night you'd need a flashlight. And all the tract models were the same way. *Caveat Emptor.*

Switches and Stairs

When the builder isn't keeping on top of things and/or is one of those who just doesn't care, you might find a switch right behind the stair railing. This is poor workmanship. (And if your house is this way after it's built, be sure it gets fixed.)

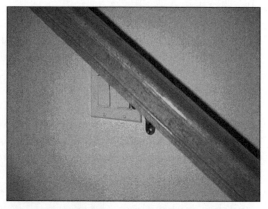

Lights for stairs use 3-way switches with one switch at the top and one at

A light switch behind the hand rail. This happened when a don't-give-a-damn builder hired an electrician with the same attitude.

the bottom of the stairs. If the stair light is also used for illumination of an area at the top of the stairs (a landing, hall, or bridge), it's convenient to have an additional switch so you don't have to walk around in the dark after coming up the stairs. Arrangements with three switches, where any one of them can control the lights, are called 4-way switches.

Switches and Doors

Since people tend to leave them open, be careful not to have switches behind doors. Also watch out for pocket doors—switches cannot be put in the pockets and, as with double doors, the electrician may have no choice but to put the switches in inconvenient locations. Thinking ahead can avoid this problem.

Double Doors—The location of light switches is a vexing problem with double doors because, as often as not, you'll have to take several steps after entering a room to reach the switch. You may get used to it, but you'll never be happy about having the switch in an inconvenient location. Ideally, there should be a wall that can be reached by opening the door 90 degrees with the switch right at the end of the door. (Location A in the drawing.)

- The arrangement shown here was seen in models in two widely separated tracts. Double doors open into the master bedroom with one door opening against a wall that would make an ideal location for the light switch (A). Instead, the switches are located where a person has to walk completely around the other door to get to the switch (B). Not user-friendly.

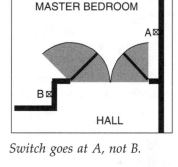

Switch goes at A, not B.

- In one expensive show home, the double doors into the master bedroom suite were removed for the show. As often happens in models, it wasn't evident that the light switch for the master bedroom and the key pad for the alarm system would be blocked by one door—or be several steps away from the other one. They should have been at the end of the door, not behind it.

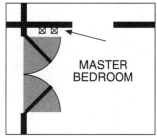

Neither switches nor key pads belong behind doors.

- It is not unusual to see expensive homes where both doors from the hall swing a full 180 degrees into the bedroom and to have the light switch behind one of the doors. There may be another set of double doors between the bedroom and the bath/vanity area. And, again, the light switch is behind one of the doors. There are better ways to do things.

If you have double doors, both of which swing 180 degrees into a bedroom or den, consider having two switches in a 3-way configuration. One switch is in the hall and the other goes inside the room at a convenient place. The switch in the hall should be next to the door that is normally used. (And this may take some intervention to make sure the electrician puts the switch where you want it and the finish carpenter mounts the doors so the one next to the switch is the one that's usually opened.) The best place for the switch inside the room will depend on the layout—putting it next to the bed is a convenient location.

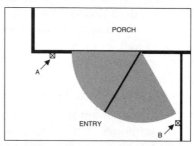

Switch goes at A, not B.

Entry Doors—Entry doors are another place where lack of foresight gets the switches in the ' wrong place. In some cases there is so much lumber around the door that there is no place left for a switch except on a far wall—or maybe it's just easier for the electrician to put it there.

Often there is a door sidelight right where you'd like to have the entry switch. There are two solutions:

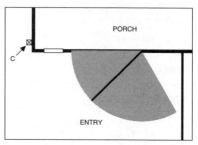

The sidelight is where the switch should go.

• Make the entry door wall wide enough that there is room for the switch (preferably next to the door) and the sidelight both. This doesn't, however, solve the sidelight security problem discussed in Chapter 11.

• Put the sidelight on the hinge side of the door which solves both the switch and the security problems.

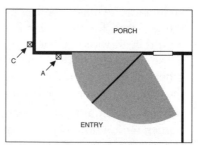

By moving the sidelight, the switch can go at A instead of C.

Ceiling Fans

Ceiling fans come in a wide range of designs and prices. If you are going to have one in your house, you'll need to do some shopping. Here are some things to think about before it's installed.

Most fans drop down from the ceiling to get the blades closer to the people they are to cool. Not all builders think about this and you'll see fans at the top of a high vaulted ceiling. You'll never feel the effects of a fan located there. Keep yours down where it'll do some good and get a larger one—52-inch diameter, not 42-inch.

But we don't always want them dropped down where we can run into them. For rooms with low ceilings (8 feet and under) get the kind of fan in which the housing for the fan motor mounts snugly against the ceiling.

Most fans are made so that lights can be added readily if they aren't included initially. A pull chain on the fan controls the speed of the fan. When lights are present, a second chain turns them on and off.

It's not absolutely necessary to have any wall switches associated with the fan unless it's so high in a room that the control chains can't be reached. However, for convenience, the usual practice is to have an on-off wall switch for the fan and, if there are lights, to have a separate switch for them. If the model house you're looking at doesn't have switches for the fan and lights, get them added. This requires three-conductor power wiring between the fan and the switches rather than the two conductors needed for the fan alone.

A common unthinking practice is to run only two wires to a single switch that controls the fan and lights at the same time. If the home user wants the fan on without the lights, or the other way around, one or both of the chains are used. When the fan is high in a room with a vaulted ceiling, it means getting up on a chair—hardly something you want—or to put a hook on a long stick that can be used to jockey the pull chains around.

Yet, that's exactly what happened to one of our sons in his new house. In the family room, the builder had planned for a simple hanging light fixture. But our son had a ceiling fan and the builder agreed to install it. The proper wiring could have been provided but no one bothered. There is now a ceiling fan with a light on it and there are only two wires from the switch to the fan.

Needless to say, when the fan is on, the light is also on and, when
someone needs a light, the fan comes on at the same time. Either
that or the long stick comes out.

Instead of having two switches on the wall, there are units with
both a fan speed control and a light dimmer that mount in place of
one wall switch! The cost is lower than it takes to mount switches
in separate boxes for the fan and the lights. In other words, it would
be less expensive to have a wall-mounted speed control for the fan
and dimmer for the lights than to have the two switches mounted a
foot apart on the wall. (An important note: don't try to use a light
dimmer to control a fan's speed. They are different beasts. You could
damage your fan motor.)

At least one manufacturer of ceiling fans has a special unit that
permits you to control the fan and lights as well as reverse the fan's
direction from a single wall-mounted controller. (But you'll need
to plan ahead. You can't install one of these fans after the house is
built unless the proper wiring is there from the start.) Still other
fans can be remotely controlled using a small hand-held unit.

Whenever it's possible to replace a light fixture in a room with a
ceiling fan at a future time, three wires should be brought from the
fixture to the light switch—not two. Also, the mounting box for the
fixture should be strong enough for a fan. These need to be put in
place when the house is built.

Ground-Fault Circuit-Interrupter Protection

In your house you have at least one outlet with two buttons la-
beled "Test" and "Reset." These are ground-fault circuit-interrupter
(GFCI) protection devices designed to protect you from electro-
cution.

Codes require that certain locations in the house be wired with
these outlets to prevent you from being shocked if you accidentally touch a water pipe and a hot wire at the same time. TV
thrillers to the contrary, you won't be electrocuted if someone
throws a hair dryer into the bathtub with you—as long as the
dryer is plugged into a bathroom outlet. Bathroom outlets have
GFCI protection that instantly turns the power off without harm-
ing you. You'll find these outlets on kitchen counters, in bathrooms, garages, and outside the house. Even if the actual
protection device itself is not present on each outlet, the National

Electrical Code requires that all the outlets in those locations be tied to a GFCI-circuit someplace.

Outlets in which the GFCI unit is an integral part of the outlet are readily identified by the two buttons on them, one marked "TEST" and the other "RESET." If you do something that trips the circuit, you have to push the reset button to get things going again. Manufacturers recommend that you test the circuit monthly by pushing the test button to see if the power goes off, then push the reset button to get back to normal.

A word of warning. The requirements for GFCI are intended for your protection. They are not inclusive. All outlets within six feet of the kitchen sink must have GFCI protection. If your toaster has a short in it and you touch it and the water faucet at the same time, you won't be electrocuted. But if you have a stove top mounted in an island or peninsula, the outlet on the island or peninsula doesn't have to be GFCI protected. Since the stove top is just as grounded as the water faucet, the outlet on the island or peninsula can kill you even if the ones around the sink don't. If your faulty toaster is plugged in on the island or peninsula and you touch it and the stove top, it's jolt time!

Don't depend on GFCI protection to keep you out of trouble. Use common sense with appliances and tools. If one has a short or a frayed cord, don't use it. You may not have protection where it's plugged in.

A GFCI outlet costs several times that of a standard one. So builders and their subcontractors try to limit the number they use. They will sometimes run wires all over the place from a GFCI outlet to avoid putting in another one.

In one house we owned the only outlet that included the protection circuitry was located in the garage and served the entire house. Thus the outlets in both bathrooms, several in the kitchen, and the outlets outside the house were all tied to this one GFCI circuit. One day I unknowingly tripped the protection circuit from an outside outlet. Because no one had explained all this to me, I spent quite a while before discovering why there was no power in the bathroom for my electric razor.

You should expect to have a GFCI outlet in each room where protection is required. When an outlet is tripped, it can be frustrating and time-consuming to try to find the GFCI protection when it's in a different part of the house—or the garage.

You'll have at least one outlet with GFCI protection in the garage. If there is more than one outlet, the others may not be protected. The one protected outlet is there for you to use for power tools. The others may be construed as being dedicated for freezers or other large appliances that, according to the National Electrical Code, do not require protection. Similarly, outside the house you may have a special "T" outlet made to handle only a single plug rather than two. This is probably for a spa. Some building inspectors will let this go by without GFCI protection. If you have one of these, treat it with special respect. NEVER use an unprotected outlet for the rotisserie motor on your barbecue, electric lawn mower or edger, or outdoor lights.

Labels are available for outlets that say they are GFCI protected. These are worthwhile. It lets you know to look at the GFCI unit as well as the circuit breaker when power is lost.

As a rule, have the builder install more rather than fewer GFCI protectors. When there is one in each room that has protected outlets, it is easier to find the one that needs resetting.

Telephone Wiring

Today there are different ways to wire houses for telephone and television than in the past. These can offer a step forward in convenience.

The local telephone company runs a cable from the street to the house for the telephones. The electrical subcontractor wires the house. The interconnection between the house wiring and the telephone company is done in the small plastic interface box found on the side of the house, usually on an outside garage wall.

It used to be that it was normal to have two lines run to houses by the telephone company. Now, because so many people have computer modems and fax machines, many telephone companies have upped this to five lines. More are available if desired but do have them put in when the lines are run initially—it'll save money and time. However, this increased number of lines in the house leads to a different concern—how to wire the house itself.

Old Two-Line Method

The older method is for the electrician to tie all the telephone jacks together (in parallel) when wiring the house. Connections between jacks are in the walls of the house. They are permanent and difficult and expensive to change. It is okay for a house with only one or two lines where future expansion or changes are not expected.

The Flexible Method

A much more adaptable method is to not tie the telephone lines together in the house but to run the two pairs of wires from each telephone jack to the telephone interface box where any telephone jack can be connected to any incoming telephone line. The connections can then be changed quickly and inexpensively at any time.

The cost for installing this method when the house is being built is more wire, which is not expensive, and using an electrician who knows what he's doing.

Jack Location

When the house is wired, you should indicate which rooms are to have telephone jacks and, if you and the builder are on top of it, the electrician will be told where the jacks are to go. If he's not told, he'll make his best guess, and this may not coincide with your needs.

In the kitchen, a wall-mounted telephone is often preferable to one that sits on a counter. At least on the wall it won't get in the way. And it's probably better if *you* decide where the portable telephone terminal is to be located.

TV Wiring

The entertainment systems in today's homes may include surround sound and multiple choices of TV signal sources (cable, roof-mounted TV antennas, satellite antennas and VCRs). These are often integrated into a single system which is both flexible and comprehensive. If you want such a system, let the builder know so that the additional coaxial and/or fiber cables can be put in place as the house is built.

Sometimes it takes a specialist to look the situation over and estimate the special wiring costs; other times the electrician can do it. It will depend on what you want. Usually you get a single TV outlet in each bedroom and in the family or great room.

The TV outlets in a house are NOT wired together. Rather, each wall connector has its own cable that is run to a central location where the cables are connected to the antenna or to the local cable company. This connection uses little boxes, called splitters, with three or more cable connectors on them. These units properly split the signals to the various TV sets, VCRs, and FM sets.

The interconnection between the cable company and the house cables usually occurs in an area such as the attic where there's protection from the weather. In some locations (like Phoenix, Arizona) the interconnections are outside the house. If there are many cables, an interconnection box someplace in the house or garage makes it convenient to change the connections to the various outlets as needed.

For your TV, wall outlets are available that have the cable jack and the power socket on a single outlet cover plate. If placed in the right place, these can help in hiding unsightly connections to the set.

The location for the TV outlet in the family room should be related to the location of the fireplace as discussed in Chapter 19.

In the master bedroom, the appropriate location for the TV outlet is one where you can watch your set while resting in bed. This suggests a location across the room from the head of the bed. Check out where you will put furniture in the room and how the TV will fit in.

As with the telephone system, check out where TV outlets are planned and, if you want something different, mark up a floor plan and let the builder know about it.

Other Electrical Considerations

- Wiring for an alarm system, security lights, and the garage door —*see* Chapter 11.
- Wiring around and under the kitchen sink—*see* Chapter 18.
- Lighting under kitchen cabinets for counter work areas—*see* Chapter 18.
- Lighting and switch locations in bathrooms—*see* Chapter 20.
- Chandelier placement for accessibility—*see* Chapter 21.
- Lights in halls for closets and thermostats—*see* Chapter 21.
- Light over the washing machine—*see* Chapter 22.
- Garage lights—*see* Chapter 22.
- Outside outlets should not be just anywhere—*see* Chapter 23.

Chapter 6

The Plumbing System

For many decades water pipe was made of iron—with its rust problems and limited life. Leaks in walls of older houses were commonplace. Then came copper pipe which, while more expensive, offered such great advantages that iron pipe became a thing of the past. Today we have several types of plastic pipe that are less expensive and potentially longer lasting than copper.

In this chapter we look at these different types of water pipes, their pros and cons, and at the two different kinds of drain pipes. Taps and faucets are generally a matter of personal preference and cost except for exterior hose bibbs where a common arrangement is anything but user-friendly.

Plumbing plans are fairly simple and straightforward unless you want something unusual like circulating hot water. There's no excuse for a layout that brings you a sudden change in shower water temperature when someone washes their hands or flushes a toilet. If you can check this in the model house, it's a good idea. You should also check that outside hose bibbs are convenient and not located where growing plants will make them hard to get to.

Types of Water Pipe

While copper has been the most commonly used pipe inside new houses, several types of plastics are becoming more popular. There has been a reluctance by the plumbing industry to make the change but now three types of plastic pipes have been approved by several building code groups, most notably the Council of American Building Officials (CABO) whose codes are the basis for many state codes.

Note, however, that a few states are slow to agree to let plastic piping be used in homes. But this is changing. You'll need to check your own state codes.

The main argument against plastic pipes is that they haven't been in use long enough to ensure that they won't cause trouble in the future—the old chicken-and-egg riddle. And, in fact, there was a serious problem with polybutylene (PB) connectors that resulted in numerous lawsuits because of damage that was done to homes by leaky PB joints. There is a consensus that these problems have been solved. Cross-linked polyethylene has been available for many years but the reluctance to use it has held back its more widespread acceptance. This is changing.

Copper

Copper is the material of choice by most plumbing subcontractors even though installing copper pipe takes more skill than plastic. Copper isn't without its problems, however. In southern California there have been reports of pin hole leaks in copper pipe presumably due to chemicals in the water. California is now starting to allow plastic piping inside residences.

Chlorinated Polyvinyl Chloride (CPVC)

CPVC is similar to PVC, the white plastic pipe used in lawn irrigation systems. Unlike PVC, CPVC doesn't soften when used for hot water. CPVC is 15 to 25 percent less expensive than copper. The installation times for CPVC and PVC are similar when done by an experienced installer.

Polybutylene (PB) and Cross-Linked Polyethylene (PEX)

Polybutylene and cross-linked polyethylene pipes are similar in that they are flexible plastics, they come in rolls, and require special fittings that are neither soldered nor cemented but are mechanical in nature. These fittings must be approved by local building officials.

The flexibility of polybutylene and cross-linked polyethylene tubes and pipes makes it possible to use them in a different way that has definite advantages, both during installation and later when in use in the home. The main water line coming into the house feeds a manifold with multiple outputs. Each output connects to an

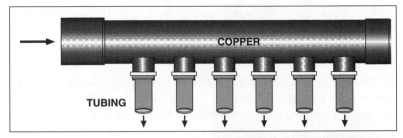

uncut piece of pipe that goes to a single outlet: a faucet, dishwasher, toilet, tub, shower, or washing machine. Joints, elbows, and couplings are not needed.

This results in lower installation times than needed for copper and a lower cost both in material and labor. Special tools and skills are needed for installation. (WIRSBO, a manufacturer of PEX, offers a twenty-five year warranty if the installation is done by a trained plumber, one year if you do it yourself.)

Because there is only one faucet per line coming from the manifold, turning on a second faucet has little impact on the flow to one already running. Thus, if you are taking a shower and someone turns on a faucet

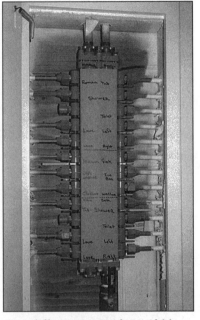

Two different types of manifolds. These offer distinct advantages over traditional methods of water distribution.

someplace else, there will be very little impact on either the amount or the temperature of your shower water.

Both PB and PEX are approved for hot as well as cold water.

Water Heaters

Be sure the water heater is suitable for your family. The tendency for builders is to use the smallest (and least expensive) heater allowed by code.

When you want hot water quickly you have two options: recirculating hot water or an auxiliary hot water tank.

Recirculating Hot Water

You are already acquainted with recirculating hot water—or did you ever wonder how you got the hot water so fast in your hotel room? The process involves doubling the amount of hot water pipe in the house. One pipe carries hot water from the heater to the faucets and the other carries the unused water back to the heater. A small pump keeps hot water circulating through the system. Both directions of pipe must be insulated if they are copper. The plastic in PB and PEX offers a certain amount of insulation in itself and additional may not be needed. WIRSBO says it's not necessary with their PEX. Check with the manufacturer of your piping and with your local building officials if you consider recirculating hot water.

There's an initial cost for putting the system into place and an ongoing cost for the electricity to run it. Both costs are small, particularly when you put a timer on the system so that it doesn't run while you sleep.

You make up at least a part of the ongoing cost since you no longer have to waste water by running it until it gets hot. If your builder will consider having it installed, find out how much it will cost and you decide if it's worthwhile for you. PB and PEX piping are attractive in recirculating hot water systems because they are easy to install, they don't have the potential noise problems of copper and the plastic pipe itself acts as an insulator.

Auxiliary Hot Water

Instant hot water for the kitchen requires available electric power under the sink and is discussed in Chapter 16.

Small heaters that mount under a vanity counter can provide hot water immediately. If you want one for a shower or tub, the heater will need to be bigger than one that furnishes water to the basins only. As with other hot water heaters, you'll need gas or electricity brought to the heater's location. The feasibility of doing this and the cost will depend on how big the heater is and on the impact it may have on the rest of the house design.

Water Pipes in Outside Walls

A good house design will minimize runs of water pipe in the outside walls because of the danger of freezing although for kitchen and vanity sinks short runs of pipe are often located there.

Outside walls in today's new houses are usually well-insulated. When water pipes are run in these walls, it is important that the pipes be kept toward the inside of the walls rather than the outside. Also, the wall insulation must be installed so that most of it is between the pipes and the outside wall and little or none is between the pipes and the inside wall. Plumbers and insulation installers understand this and normally install it this way. However, it doesn't hurt to check. (The use of toe-kick registers also helps— *see* Chapter 7)

Nail Plates

Pipes running through studs or plates are always in danger of being punctured by wallboard fasteners—nails or screws that are 1¼- to 1⅝-inches long and are driven into studs behind the wallboard. When the wallboard goes up, the installer can't see where the pipes are. Plumbers are supposed to put pieces of galvanized sheet metal, called "nail plates" or "safety plates," on the outside of the stud or plate to prevent the wallboard installer from putting a nail or screw there and inadvertently punching a hole in the pipe. The builder or his construction supervisor should check that nail plates are in place properly before the wallboard is installed, but they are sometimes missed.

Nail or safety plates protect pipes during wallboard installation—if they are not forgotten.

Besides the obvious problem when an immediate leak occurs, improperly placed or missing nail plates can result in subtle problems that may not be seen for several years. Small leaks in drain lines are not always self evident and the dampness may be contained

inside walls where it can rot
the wood without being seen.
There is also the possibility
that a wallboard screw will
penetrate a water or drain
pipe but the screw itself plugs
the hole—until it rusts out a
few years later. If you can,
double check that nail plates
are in place to protect *all*
pipes before the wallboard is
installed. The builder's warranty for things like this lasts
only one or two years.

Oops! A nail plate is missing!

Noisy Pipes

You can reduce water and drain pipe noise by paying attention
to how the pipes are installed. As an experiment when you visit
model homes, have your spouse turn on a faucet in one end of the
house and see if you can hear the running water in the other. Some
houses are significantly better than others in this respect.

Water running through a pipe is noisier when the pipe touches
wood because the wood can act as a sounding board. Pipes should
be kept away from joists and studs. Plastic piping used in water
distribution systems has additional advantages here. First, most
noise originates where the direction of flow changes abruptly and
this doesn't happen with plastic in normal installations. And, copper will carry noise along the pipe more readily than will plastic.

Drain water falling from the second floor of a house makes noticeably more noise in plastic than in iron pipes. If iron pipes are
wrapped with insulation, the noise from the second-floor drain
pipes will be just about eliminated. However, iron pipe rusts, so
check its expected life in your area before having it installed. For
either iron or plastic, wrapping with fiberglass insulation will help
deaden the noise.

Hose Bibbs

You'll want hose bibbs on the outside of your house. Most builders put in two: one in front and one in back. For most of us this isn't
enough. If you have an RV pad, you should have a hose bibb there.

In any case, an additional hose bibb on at least one side of the house is useful.

If you need hot water to wash your car, have a bibb installed in the garage, connected to both hot and cold water. (This works well with the utility tub in the garage as discussed in Chapter 19.) Be sure you take a hard look at the type of faucet that is used. The common faucet used for mixing hot and cold water restricts the amount of water that can be passed. This is no good when a high volume of water at full pressure is needed.

Make it known, either in writing or on a drawing, where you want the hose bibbs. Allow the plumber some freedom to minimize the cost of putting them in place, but be sure your wants are known.

In areas where there is danger of water freezing in hose bibbs, plumbing codes require a means of draining the water from the bibbs in the winter. Two methods are permitted in the codes: a regular hose bibb with a stop-and-waste valve or a frost-proof hose bibb.

Stop-and-Waste Valves

Stop-and-waste valves are installed in an accessible heated location where there is no danger of freezing. The "stop" part is simply a turn-off valve. The "waste" part is a plug on the valve which can be removed to allow water to drain.

The hose bibb is connected to the stop-and-waste valve with pipe that may be exposed to freezing conditions. In the autumn the home user must:

1. Turn off the water at the stop-and-waste valve.
2. Open the hose bibb and take care of any hoses which may be connected.
3. If the bibb and pipe don't drain by themselves, the waste plug must be removed to let the water out.

In the spring the above process must be reversed.

Unfortunately, not all plumbers follow the code carefully and not all inspectors see that they do. In far too many cases the stop-and-waste valve is placed where it is not readily accessible, making it virtually useless. If your builder or plumber insists that frost-proof hose bibbs are not a good idea, then you should insist that he put the stop-and-waste valves where you can get to them easily and that pipes drain by simply opening the bibb.

Frost-Proof Hose Bibbs

The frost-proof bibb includes a piece of copper pipe as a part of the bibb itself. This pipe extends the valve part of the bibb back into a part of the house that is heated and where there is no danger of freezing. When it is turned off, water drains from the exposed part of the hose bibb and no damage occurs.

There is a precaution, however. Sometimes hoses with closed nozzles are left connected so that the water cannot drain. This has happened often enough that plumbers in some localities have an aversion to using these bibbs at all. Properly used, they effectively eliminate problems with freezing.

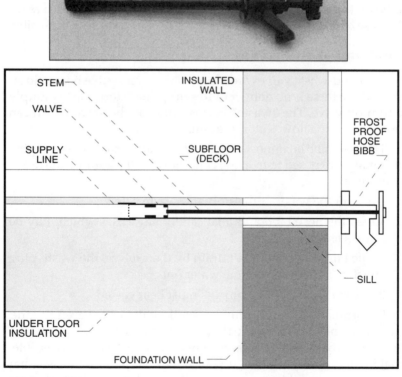

A frost-proof hose bibb.

Chapter 7

The HVAC System (Heating, Ventilation and Air Conditioning)

The HVAC system in a new house today is much more than just a way to control the temperature. It, and how the house is built, determine energy efficiency, comfort, and the size of your heating bills.

The mechanical subcontractor is responsible for the HVAC system.

When you visit model homes, take a look at the type of heating system that is used. Most will be forced air with electricity, gas, or oil as the energy source but you may find something different such as hot water baseboard, radiant floor heating, electric baseboard, or wall heaters. These non-forced air systems can be advantageous, particularly when most of the time you want to heat only a small part of the house. Their disadvantage is that they cannot be integrated with an air conditioning system. Some comments about these are followed by a more detailed discussion of the forced air systems found in most new homes.

Non-Forced-Air Systems

Since only forced-air systems can be used for cooling, manufacturers of non-forced systems will tell you how to have air conditioning at a lesser cost than a full-blown forced-air heating and cooling system. If you see one of these arrangements, consider it carefully for convenience, comfort, cost, and resaleability. These systems do a good job of heating on a room-by-room basis; if this is important to you, one of them may be the way to go. (As explained below, you can accomplish much the same thing with a forced air system using "zoning.")

Hot Water Baseboard—With a hot water system, the "furnace" is a water heater from which hot water is piped to the various rooms in the house. Baseboard heaters are basically fins attached to pipes that are stretched along the baseboards on several sides of the room. Thermostats control small pumps and each room can have its own thermostat so that only the rooms in use need to be heated. The basic energy source is usually gas or oil.

Radiant Floor Heat—These systems use hot water in plastic pipes laid in a special solid matrix under the floor covering. In other ways they are similar to a hot water baseboard system. A claim is made that this system heats rooms more evenly than forced air systems. This is probably true. However, it also radiates heat under furniture where it is not needed or wanted.

A potentially more severe problem with this system is repair. If a pipe springs a leak (and it isn't clear why it ever should), it is necessary to take up the floor covering (rug, hardwood, linoleum, or tile) to expose the matrix which is then dug out to get to the pipe. As was discussed in Chapter 6, plastic pipe, such as polybutylene, cannot be cemented. To repair it, enough has to be exposed to permit two mechanical connectors and a replacement piece of pipe to be joined together. While this shouldn't have to be done very often, even once in the lifetime of the system is a significantly bigger undertaking than repair of any of the other systems.

Electric Baseboard—As with hot-water systems, each room has its own thermostat that controls the electric baseboard elements in it.

Wall Heaters—With these systems, heaters are built into the wall. Fans force the heat out into the room. Depending on the system, each room may have its own heater or it may depend on air movement to heat several rooms from a centrally located heater.

Forced Air Systems

In forced air systems, air is heated or cooled and forced under pressure through a network of ducts to outlets or registers in various parts of the house. Usually there will be one such system per house, but in larger homes there may be more.

The basic energy source for forced air systems may be electricity, gas, or oil. Electric systems usually use a heat pump because it is more cost-effective than heating the air with electric heating elements.

Cooling a house involves collecting warm air using the duct work, cooling it and redistributing it. The cooling is done using refrigeration techniques that take the heat out of the air in the house and send it outside. The machine that does this may be an air conditioner or a heat pump.

Heating is similar. The air is heated by a burner in a gas or oil system or by the heat pump in the electric system. Note that the heat pump is used for both heating and cooling. This is accomplished by reversing the refrigeration process. For cooling it moves the heat from inside the house to the outside and for heating it takes heat from outside and moves it inside.

Which heating and cooling approach is best for you will depend on the climate in your area, whether cooling is important to you, and on the relative costs of the available sources of energy. Your local heating and air conditioning equipment supplier can best advise you. (Don't depend on getting straight answers from energy suppliers—they have their own agendas.)

Of particular interest to us here are several aspects of the design of forced-air systems which impact directly on their effectiveness in making our homes comfortable places to live.

System Design

Ducts—In one-story houses with crawl spaces, the ducts are usually run under the house and the registers are in the floor. With concrete slab floors, the ducts, and often the furnace, are in the attic and the heating registers are in the ceiling or in walls close to the ceiling. With two stories, the registers may be in the floor or in the ceiling.

The forced-air heating system in our one-story house was designed by the mechanical subcontractor—the builder simply hired him and depended on him to do the job right. When the heating season came we found the bedroom wing of the house was too warm and the living areas weren't warm enough. We closed the floor registers in the bedrooms to balance the system. In trying to understand why this was necessary, I found that the two areas have about the same number of square feet and that the basic design put about the same amount of hot air into each of them. However, because the living area of the house has a larger proportion of glass and outside wall than the bedroom wing, it has significantly more heat loss. The basic heating design was not done right.

If you get the chance to have a knowledgeable person review the HVAC plans before they are implemented you may be able to avoid problems arising from a bad design. It may not always be possible to find someone to do this for you, but it is good insurance if you can.

Type of Register—Two different register types are in common use; what you have will depend on where you live. Floor registers that mount flush with the floor, ceiling, or wall are used almost exclusively in the west.

In other places registers may be a type that mounts against a wall just above the floor. They look similar to a baseboard heater.

Their advantages are that they are not as restrictive of furniture place-ment as floor registers and they are not a problem for foot traffic. Their disadvantage is that they are not as easy to clean as a flat register. But don't use them (or any other register) in outside walls where they would occupy space that is normally used by insulation. This will result in increased heat loss from the register area. (This will be noticeable during the winter as cold air that comes from the register when the furnace is off.)

This register directs the heat into the room instead of up.

Registers should not go in outside walls like this one. Insulation belongs there.

Register Location—It has been standard practice forever to put heating registers around the outside walls under windows. This was done to offset the cold air that came off windows in the winter.

When you have high-quality double-pane windows this problem is greatly reduced and the old rule is no longer as iron clad as it was. Take a look at the model or at the mechanical subcontractor's plans before work is started on your own home. If you want to put a piece of furniture in front of a window see if the register can't be moved along the wall to where it won't interfere with furniture placement. (Make the arrangements to have this done before work starts on the house—it may impact on the cost of the HVAC system and on the price of the house.)

B ut don't do what I saw in a tract in York, Pennsylvania where the builder **was** thinking ahead and had moved the register to a different wall away from the window. Which was great except it was the only wall where the bed could go and there it was—right under the bed in the furnished model!

You will see cold air returns in upstairs hallways and bridges—areas where people could be walking in bare feet. These are large registers and clearly the builder wasn't thinking when this happens. Wooden grills on the registers are bad enough but metal ones are worse. With a little forethought they can be mounted in an adjacent wall.

Registers are not designed to take foot traffic nor are feet designed to walk on registers. When heating registers are put in front

Registers in front of sliding doors or in other traffic areas are an unnecessary pain in the foot.

of sliding doors the builder must coordinate the work of the mechanical sub and the door installer, otherwise there's a good chance
the register will be installed in front of the sliding part of the door.
When it ends up where it will be walked on, the builder wasn't
doing his job.

In kitchens and bathrooms with cabinets, the registers may be
placed in the toe kick region at the base of the cabinets. This has
several aesthetic and functional advantages over floor registers:

- Nothing dropped on the floor will get into the register or duct
  work. (Ever go searching for a diamond ring down a heating
  duct?)

- Mopping a floor can be done without concern that mop water
  will get into the register or ducts.

- The registers cannot be walked on, a relief for people who may
  go around their kitchens or bathrooms in bare feet.

- If you live where there is a potential of pipes freezing in outside
  walls, they are protected by this register arrangement because
  the inside of the cabinet and the pipes are kept warm by the duct
  under it.

*A toe-kick register
is a user-friendly
improvement over
one in the floor.*

I n yet another example of a builder not paying attention, the cabinet installer forgot to cut the toe-kick register into the base of the kitchen cabinet in our house. When cold weather came, we noticed that the kitchen was the coldest room in the house and there was no way enough heat could be forced in to make it comfortable without overheating everything else.

Then one day, while looking at houses, I noticed a register in the toe space under a kitchen sink. Sure enough, back home I found the cabinet area under the sink was toasty warm but the heat had no way to get out. After a discussion with the builder, the cabinet installer was called back to finish his job, the register was put into place, and the problem was solved. (Where was the builder to let this happen? And what would have happened if I hadn't been looking at houses? **Caveat emptor.**)

Builders' Dirt in the Floor Registers—There is a further problem with floor registers—construction dirt and debris. The mechanical subcontractor installs the heating ducts at a time when much of the house is unfinished. To prevent rain and dirt from getting into the ducts, a good mechanical subcontractor seals them immediately after installation. When the outer shell of the house is completed, unless forbidden by code, the builder takes the covers off the ducts, leaving them wide open, and turns the furnace on to dry out the house. Then come the wallboard, floor coverings, finish carpentering, interior finishing, wallpapering and painting. From all of this a significant amount of debris, dirt, and dust fall into the open boots and lies there in wait for you, the new owner, when you move in.

Debris from one uncovered register.

The worst is the dust from wallboard work. This is very fine and will be in the house forever unless you get a special set of furnace filters. We were in our new house for almost two years with this coating of fine dust always present. Different filters caught most of it.

When the builder is going to use the furnace for warming and drying the house during the latter phases of construction, he may do one of several things:

1. Nothing. This is the easy way out.

2. Clean out the ducts before turning the property over to the buyer. This won't get it all because it simply isn't possible to get high-power vacuum hoses through the whole ducting system. (I got a whole handful of debris from a duct that the builder paid to have vacuumed—but never checked to see if it was done right.)

3. Not open the ducts until construction is complete. With this approach a portable propane heater is used to heat and dry out the house. One builder I know also uses a dehumidifier to remove the moisture.

4. Buy a set of special air filters to go over the open registers that lets the heat out but does not let debris into the duct. These filters are removed and thrown away after the construction work has been completed and the floor covering installed. They are available from Quality Air Systems, 4820 115th Avenue, Clear Lake, Minn. 55319, telephone (612) 743-2627.

5. Put caps over the ends of the ducts or the boots whenever workers are in the house and run the furnace only when they are gone. If you are involved when the house is under construction, you may be able to arrange with the builder to do this yourself. It's one way to know it's done right.

Collapsed Ducts—A boot is a piece of metal shaped to connect to the round heating duct on one end and to hold the register at the other. In one house, when I put my arm down to check the debris in the duct, I could hardly get my hand into the 4" duct. The flexible duct had been bent so much where it connected to the boot that it had collapsed, restricting the size of the opening. Whoever bought that house would have had a problem with his heating system. (I told the builder about it, so it should have been fixed before the new owners moved in.)

Note that collapsed ducts are not unusual. They shouldn't happen but they do. When you find one, there should be no problem having it fixed. And it's a good idea to check every one in your new home before moving in.

Master Bedroom Air Return—Consider having an air return installed in the master bedroom suite. This serves two purposes. First, when the door of the suite is closed, the flow of air from the registers to the return will not be disrupted. It is not unusual to find that a master bedroom is comfortable when the hall door is open but gets too warm or cool, depending on the time of year, when the door is closed. With a return in the suite, this won't happen.

Second, it keeps a closed door, especially a double door, from rattling every time the furnace goes on or off. When the door is closed and there is no place for the air from the registers to go, the pressure will shove the door hard against the jam. When the furnace fan stops, the pressure is released. The result is a door that can make disturbing noises in the middle of the night. This can be helped by putting several little rubber stick-on bumpers on the door jam, the same type that are used on cabinets to keep them from banging. (This isn't a bad idea in any case; it makes it much easier to close a door quietly.)

House Design versus HVAC Design—If yours is a big house and the heating and cooling equipment is far from some of the rooms there may be no good way to get enough hot or cold air to them. A supplementary system for these rooms is often a good idea but many designers and builders don't know or care about this; they leave it up to the HVAC contractor whose hands are tied by the time he is brought in. If you have any doubts, you and your designer or builder should consult an HVAC contractor before the plans are finalized.

Zoning—Even with forced-air systems, it is possible to break the house into zones that are heated and cooled independently by using separate thermostats for each zone and having electrically controlled dampers in the ducts. A programmable control unit controls the temperatures by zone. You can change the program according to the time of day or even the day of the week!

The usual zoning arrangement is to control the upstairs and downstairs of a house separately. This leads to more comfort since upstairs get hotter both from the natural tendency for warm air to rise and from the sun beating on the roof. A zoning control system is more efficient and less costly than the alternative of two separate systems.

Zoning can also be applied to individual rooms, such as solariums, that have special heating and cooling requirements.

High-Efficiency Gas Furnaces—High-efficiency furnaces extract up to 90 percent of the heating potential from the gas that is burned. But there is a precaution. In these units there is not enough heat left in the vent to keep steam in the exhaust from condensing. This results in water collecting at the bottom of the vent which must be gotten rid of someplace. If planned for ahead of time, this need not result in an eyesore or an inconveniently placed drain pipe. If you are going to have one of these furnaces, talk to the builder and the HVAC subcontractor before everything is literally poured in concrete.

Heat Pump and Air Conditioner Considerations

If yours is a forced-air system and you want an air conditioner with it, either initially or in the future, or if you are going to use a heat pump, it's a good idea to pick the location of the external unit before the house is built. Sometimes the only available space, by the time the system is actually installed, makes it an eyesore. And have the HVAC sub look at the plans to see what problems there may be in getting the wires and tubing between the outside and the inside units. Planning ahead can avoid headaches.

"Air Conditioner Ready"—Some house ads read "air conditioner ready." This can mean one of several things: the wiring and piping are in place ready for the air conditioner, the wiring is in place, or only that the furnace has space for adding the expander coils. Adding wiring, once the house is finished, is a fairly expensive proposition and may involve running the wiring in a conduit on the outside of the house. It's worthwhile, if there's a possibility of adding an air conditioner, to have the wiring done when the house is built. Depending on the house design, it may also be worthwhile to make the openings in the foundation or elsewhere for the pipes that go between the furnace location and the outside compressor.

Heat Pumps—In practice, heat pumps are more expensive to buy and less expensive to operate than separate heating and cooling units. In climates where the outside air temperature gets too low for efficient winter pumping, two options are available: 1) use a gas, oil, or electric furnace as a backup or 2) use the ground as the source of heat. The latter arrangement is more expensive to install, but doesn't require as much backup heating.

People who use heat pumps report another concern—houses with pumps respond to temperature changes slowly. This is because the amount of heat they can pump is limited. With modern, well-insulated houses, heat pumps keep up with day-to-day temperature changes quite well. But if you are away from home with the heat pump turned down or off, you can expect it will take longer to get the house back to normal temperature than it would with a separate furnace or air conditioner.

Thermostats

Several types of thermostats are available. The simplest has a single setting for controlling the heater. You turn it down at night and you turn it up in the morning. At the other extreme is a computer-controlled unit that resets the operating temperature several times a day with different settings for weekdays and weekends. As a minimum, it is a good idea is to use a unit that includes a clock that lowers and raises the thermostat setting once a day at preset times. This will result in a more comfortable house and in savings in heating and air-conditioning operating costs. These are called "setback" thermostats and are an energy code requirement in some states.

If you have trouble setting up your VCR, you should probably avoid the digital units, particularly those that include a lot of functions—they can be tedious to set. It's worthwhile to visit your local heating supply company or home improvement store to get a look at different types of thermostats.

Location—The thermostat can control the temperature only at one location, so it should be located where you want to control the temperature most closely. It should not be near a heat register because the furnace will cycle on and off more often than you would want. Since most heat registers are around outside walls, putting the thermostat in the center of the house usually works best. Normal practice is to put them more or less in the flow of air toward a return register.

Thermostats should be installed where average people can see the settings comfortably with little stooping or standing on their toes.

Seeing Your Thermostat—For a mechanically adjustable thermostat, insist on one that lets you easily see where it is set and then coordinate the locations of thermostat and lights. If you choose a unit with digital readouts, don't forget that you will need light to see the digital display and you will also need light to see the keyboard when the thermostat is being programmed.

Let the builder know the type of thermostat you want and that you want it to be lit, either internally or by a nearby light.

Chapter 8

Fireplaces and Stoves

Except in rural areas you don't see stacks of wood piled around the house as winter sets in. Smoke from chimneys is mostly something from pictures and paintings of the past. Stoves are still used as room heaters but they burn pellets or gas instead of wood or coal. And fireplaces used for heating don't burn wood but gas, with efficiencies that are surprisingly close to those of central heating systems.

While stoves and many fireplaces are primarily for heating, some fireplaces are used mostly for the ambiance they provide. Three types of fuel are common: wood and gas in both stoves and fireplaces and pellets in stoves. The efficiencies—how much of the potential heat in the fuel is used to heat the room—vary widely.

State code bodies also vary widely in what they think about fireplaces and stoves, resulting in significant differences among the codes.

Since one of the aims of state code officials is to save energy, one of their concerns is the efficiency of the designs used in stoves and fireplaces. For several reasons discussed in this chapter, the most fuel-efficient stoves and fireplaces are those whose fireboxes are sealed with both the combustion air and the exhaust being completely separated from the air in the house. (Combustion air is the air the fire uses to burn.)

Stoves which use fans and fuel augers are not included in the following discussions. These stoves are found in older homes where there is a need for a unit which can deliver enough heat for several rooms. They burn continuously and are often thermostatically controlled. None of them has been seen in new construction.

Here we are concerned with fireplaces and stoves which depend on convection to get the the burned gases outside.

Pilot Lights

Pilot lights are used so that gas stoves and fireplaces can be turned on and off with a switch—matches or lighters aren't needed. The pilot is an integral part of the design because the heat from the pilot light is used to thermoelectrically generate the power needed to turn the gas on and off from the switch. This ability to operate without external electricity is one of the attractive features of these units—they don't depend on electric power and will continue to operate during power outages.

The down side is that the pilot burns all the time, even in the summer when it isn't needed or wanted. For this reason some state (California is one) codes do not allow them in stoves or fireplaces.

Fireplaces Used for Ambiance

Fireplaces designed for wood or gas logs are used to provide the ambiance that goes with flames flickering among the logs. This is not to say that they don't generate heat but rather to note that they are not efficient ways to get heat from the fuel.

If you live where a continuously burning pilot is forbidden and want a fireplace, it has to be wood-burning or gas logs. (If you find

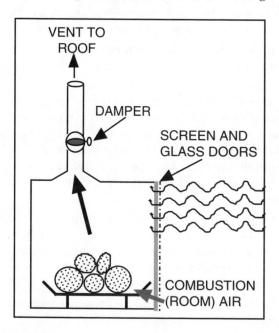

Wood-burning fireplaces need a damper to prevent loss of heated room air.

a gas fireplace that is sealed but does not use a pilot light, get it. The rest of the argument here is then moot.)

One of the problems with wood-burning fireplaces or gas logs is the damper. When a fire is burning, a closed damper will force smoke or burned gas back into the room. With the gas turned on but not burning, a closed damper is a recipe for an explosion. Even without codes, such a possible scenario should give you second thoughts.

And that is the reason a number of state codes mandate that the damper be blocked permanently open when gas is available at the fireplace. But note that this open damper lets heated room air continuously escape up the flue whether the fire is lit or not.

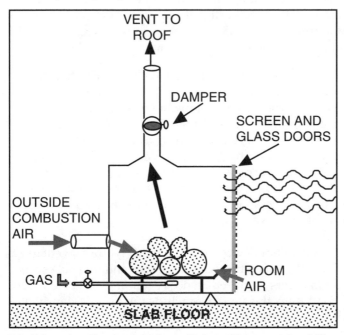

Wood burning fireplace with gas lighter. Dampers should be blocked open when gas is used and several state codes require this.

The open damper is a relatively recent requirement. Our southern California house had a gas lighter but the damper wasn't blocked. We tried to remember to close it when the fire wasn't burning—and to open it before turning the gas on. We had no problems, but the changed codes are obviously appropriate.

If you want a wood-burning fireplace or gas logs, the blocked damper will cost you money. Even if there is gas in the house for heating or cooking, don't be in a hurry to have it piped to the fireplace either for gas logs or for use as a gas lighter because that'll mean an open damper forever.

Efficiency

The efficiency of a heating device is the ratio of the actual heat generated to the available heat from the fuel. Efficiencies are averaged over a year's time using standardized methods. Whatever heat potential isn't used goes up the vent pipe and is lost. We see efficiency ratings on gas heating units—water heaters, stoves, furnaces, and fireplaces. Ratings typically run around 85 to 90 percent for furnaces and upward from 55 percent for fireplaces and stoves. Efficiency ratings for oil-burning devices are comparable.

When I was growing up, we used a portable gas heater to take the chill off of a room on cold winter mornings. Gas was piped to the rooms where it might be used and the heater was connected when it was wanted. This device was 100 percent efficient—since there was no vent, all of the heat from the burning gas stayed in the room. Such devices are no longer used because they are dangerous: it is easy to catch something on fire if it gets too close to the open flame and, if misadjusted, can lead to the formation of carbon monoxide. But they were 100 percent efficient!

There are equivalent devices available today—the unvented gas fireplace. Burned gas is returned to the room and no heat escapes. To eliminate dangers from reduced oxygen and possible carbon monoxide formation, this fireplace includes an oxygen-deprivation sensor that turns off the gas if there is a danger.

There are two potential problems with the unvented fireplace:

1. The combustion products from the burning gas go right back into the room the same as the old-time room heater. These will be primarily carbon dioxide and water vapor but that's not all— there are other products coming from minute amounts of impurities in the gas. These are taken outside with vented fireplaces and stoves but are dumped back into your house with the unvented units. They may not be dangerous—or presumably codes wouldn't allow them. But they do depend on the purity

of the gas and accu-
rate adjustment to
keep unwanted
c o m b u s t i o n
particles(greasy
soot) from messing
up your drapes, fur-
niture and carpet.

2. In a tightly sealed
house there may be
no way for replace-
ment oxygen to get
in, which means the
fireplace will turn
itself off after it has
been burning long
enough to affect the
oxygen level. While
safe, a fireplace that
turns itself off may
not be to your lik-

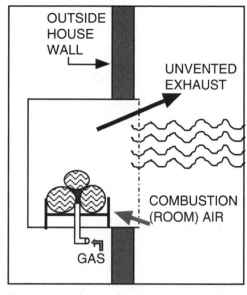

OUTSIDE HOUSE WALL

UNVENTED EXHAUST

COMBUSTION (ROOM) AIR

GAS

The no-vent fireplace exhausts burned gases back into the room.

ing. And, of course, the safety of the device depends totally on
the oxygen-deprivation sensor *never* failing.

There is yet another side to this story. Neighbors unwittingly
bought a house with two of these no-vent fireplaces. After
they moved in they realized what they had. Now, three years
later, they haven't turned them on because they are well aware
of the potential problems. If you get a house with such a fireplace, be sure you understand that you may be making your
house harder to sell when the time comes because many people
simply won't want any part of a no-vent fireplace.

Direct Vent versus Top Vent

The terms direct vent and top vent are confusing in today's stoves
and fireplaces. When the firebox is not sealed, the combustion air
may come from anyplace but some of it *will* come from the room.
Like old-time stoves, the exhaust vent is usually on top of the unit,
hence the name top vent. But the exhaust vent can be on the back
or side of the unit just as well and combustion air can come primarily from a pipe to the outside (usually a crawl space) or it can all

come from the room. This type of venting is properly called "type B." You'll also still see and hear them called "top vent."

Direct vent is the term used to describe a sealed fire box where none of the combustion air comes from inside the house. A common version has two concentric pipes (one inside the other) to get combustion air in and exhaust gases out. These pipes may go straight out the back of the unit making a short direct run through an outside wall. Today's stoves and fireplaces with sealed fireboxes, even though still called direct vent, come in several different arrangements and are not limited to running concentric pipes straight out the back.

Top Vent

There are two problems with top-vent units that direct vents do not have:

1) They cannot be used in a tightly sealed house.

2) They lose heated room air continuously whether lit or not.

For convection to work with top-vent units there must be a supply of air coming into the firebox: air that is heated in the combustion process and then goes up the vent. If there is no "spare" air, there is no convection and the fireplace or stove doesn't work. In a tightly sealed house this may be the case.

The fireplace could get its combustion air from outside, say, the crawl space, and then it would always have a supply of air for the convection process to work. But, unless it's a sealed firebox—in which case it would be a direct-vent unit—it is not sealed off from room air. Now, when there is no fire, the outside air can come directly through the firebox and into the room—and who wants outside air coming out of their unlit fireplace?

This problem is exacerbated when a kitchen exhaust fan is working. It, too, needs to get its air from someplace and it'll pull it down the stove or fireplace vent as well as from anyplace else that isn't sealed.

We've seen that a top-vent unit won't work in a tightly sealed house but, if it is not well-sealed, it opens the door for the other problem with top-vent units. There is absolutely nothing to prevent heated room air from escaping—continuously!

The top-vent gas fireplace in my house gets its air through a four-inch duct under the firebox. The duct could have been

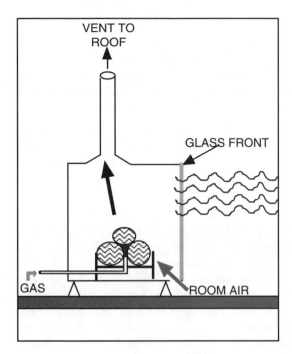

VENT TO
ROOF

GLASS FRONT

GAS

ROOM AIR

Heated room air is continuously lost in a top-vent (type-B vent) gas fireplace or stove.

connected to get the combustion air from the crawl space under the house but it wasn't. To see just how much heated room air might be going out through the fireplace, I put a lighted butane BBQ lighter in front of the duct opening. (The fireplace was off.) There was a good flow of air—the flame was bent side-ways. When the same thing was done with a breeze blowing outside (which causes suction), the rush of air into the duct blew out the butane lighter instantly. And this is all my heated room air! And it goes on and on! The gas company would love it if they weren't so energy conscious.

This is not a way to measure how many BTUs the fireplace loses but it does show that this isn't the way to go. It's similar to leaving a four-inch hole in the ceiling. And note that **any** flue in an unsealed firebox without a closed damper—whether stove, wood-burning fireplace or gas fireplace—will do exactly the same thing.

The rated efficiencies of top-vent fireplaces and stoves do not take this loss of heated room air into account. But it's there, it's real, and it costs money.

Direct Vent

Direct-vent fireplaces and stoves are sealed so that the only interaction between room air and the firebox is the exchange of heat through the glass front. This unit should be the choice for a new home whether the house is tightly sealed or not—unless you live in California or some other place that doesn't allow pilot lights.

The direct-vent stove or fireplace takes the convection consideration one step further. The rising of the heated air in one pipe pulls outside air into the fire box through the other one. Anything that restricts the air flow will lower the efficiency of the unit so it is important that the pipes be straight.

The original direct-vent fireplace or stove used concentric pipes and had to be mounted against an outside wall. But outside walls aren't always where people want their fireplace or stove. Now stoves and fireplaces are available where the concentric pipe connection is on the top of the unit it so can be run straight up through the roof. Other units have separate connections for the air intake and the exhaust—again on the top of the unit. With these arrangements there are no restrictions on the location of a direct-vent stove or fireplace—except, maybe, one due to the length of the exhaust duct. You'll have to check that out with the manufacturer.

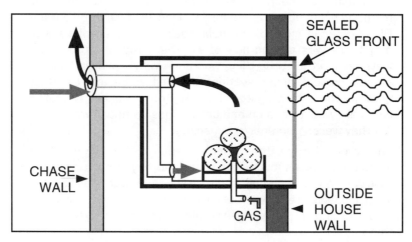

There is no loss of heated room air with a direct-vent gas fireplace.

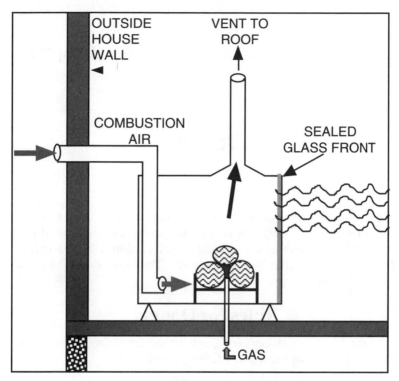

A direct-vent gas fireplace with separate intake and exhaust vents.

Insulation Around Fireplaces

Neither the masonry nor the metal used in fire boxes is a good insulator. To avoid heat loss from fireplaces on outside walls, the back sides should be enclosed in an insulated chase.

Two-Sided Fireplaces

A two-sided fireplace between a family and living room doesn't raise a privacy issue. But between living room or family room and the master bedroom, it is a whole different situation. I've seen homes built with these arrangements and a large part of the master bedroom is clearly visible from the living room. The users of these homes will have to provide some kind of screen in front of their fireplace if they want to be alone. You may want to have back-to-back fireplaces in the two rooms instead.

Chapter 9

Doors

One of the more obvious signs of a poor house design is to find doors that are used inappropriately: hinged doors that should be pocket doors, doors that cover light switches, doors that bang into other doors, doors that block doorways. The ways door *should* be used are discussed here.

Rules for Doors

Unsuitable use of doors is one of the most irritating, least excusable, and *potentially hazardous* of design sins.

Here are Ferguson's three door rules:

(1) A door should never open across another door of any type.

(2) Any door that may normally be left open should never be in the way when it *is* open.

(3) A door should never swing where someone is apt to be standing or sitting.

Some door arrangements can violate more than one rule. In the following discussions you can see again that it's just common sense. If one of these arrangements is in the plan or house you want, see if you can't do something to improve the situation before you make a commitment. Easy fixes are not always possible.

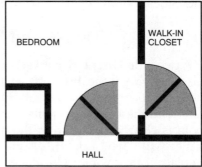

A hazardous violation of Rule 1.

Rule 1. Doors Opening Across Doors

Consider the figure on the previous page. If you're coming out of the walk-in closet when someone is coming into the bedroom, you and the door will be in the same place at the same time. A door in the face can be much more than just irritating. Avoid combinations like these. (In practice, if you had a house with this in it, you could swing the door from the hall on the other side which would improve the situation—but not cure it.)

Extend this rule to include closet, cabinet, and appliance doors as well. While these doors are normally closed, having to close a room door before opening a closet or an appliance door just doesn't make good sense.

In one custom house I visited, the door from the master bedroom into the bathroom opens right across the door into the room where the toilet is. So to get to the toilet, you go from the bedroom into the bathroom, then you must close the door you just came through so you can open the door to get to the toilet. Wanna try this in the middle of the night? (The last drawing in this chapter shows the floor plan.)

Rule 2. Open Doors Should Not Be in the Way

Perhaps the most obvious application of this rule is to walk-in closets where you shouldn't have to close the door to get at your clothes. The photo shows another one: you have to close the door into this utility room to be able to use the closet.

Note that hinging the door on the other side would improve things considerably because you could then leave the door open. Actually, the plans show this but the carpenter did it his way. You have to watch everybody!

This door arrangement violates both Rules 1 and 2.

Still another application of this rule deals with external hinged doors. Most of us like to be able to open these doors for a little fresh air when the weather's nice and will install a screen door to keep the bugs out. Yet if there has been no forethought, when the door is left open it will be in the way. (This is discussed in more detail below.)

Rule 3. Watch Where the Door Swings

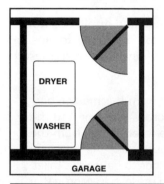

In far too many cases you'll see doors that are dangerous to use because there's apt to be someone on the other side. The example for Rule 1 is one of those cases. The drawings show a couple more.

Anyone doing the washing and leaning over the washer is vulnerable to being hit by the door from the garage. What to do about this is discussed in Chapter 22.

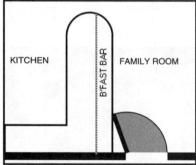

In the other example, the usefulness of the breakfast bar is degraded by the door. A sliding door would be far better.

Entry doors that open across stairs will block the stairs as long as the door is open. This is particularly annoying when you come in the front door with something to carry upstairs.

Two Rule 3 no-nos.

Types of Doors

The different types of doors used in today's houses are:

1. Hinged doors. The most common type, they are used everywhere, including places where they shouldn't.
2. Sliding glass doors. Used for entering and exiting the house.
3. Sliding wooden doors. Used as closet doors. Sometimes have mirrors.
4. Pocket doors. Used as interior doors where hinged doors would get in the way.

5. Bifold doors. Used primarily for closets. May be louvered. The doors themselves may be mirrors.

Exterior Doors

Entry doors are usually decorative and wood is the most popular material. Wood is disadvantageous in three ways: it can warp, finishes deteriorate rapidly when exposed to weather, and it is not a good thermal insulator. If appearance is not a prime consideration, you should consider exterior metal doors. These are a metal sheath over a foam interior. They transmit less heat than wooden doors and have better lasting finishes. Vinyl doors that simulate wood are available and these are also foam-filled.

Note that, unlike interior doors discussed below, exterior doors are never hollow core for two reasons:

1. They provide better insulation than do hollow-core doors.

2. Building codes require that solid-core doors be used between the house and the garage because they are better as fire blocks.

Exterior doors that swing out are potential security problems because the hinge pins are on the outside where they can be easily removed. Special hinges are available to overcome the security problem but, if you want a screen with these doors, it will have to swing inside. And there's no place for a storm door.

Accessible hinge pins make it easy to get into a house.

Fire officials deplore doors that swing out and in some jurisdictions they are forbidden by code in new homes. When firemen have to get into a burning house with locked doors, they must break in. It's much easier to push a door in that is hinged on the inside.

It's just better to avoid doors that swing out.

Even the best of insulated exterior sliding glass doors passes much more heat than an insulated door without glass. Given the large amount of glass in sliding doors, they will always be a major source of heat loss. Use them sparingly, even those with the best insulation properties.

Locks for sliding doors that provide a reasonable amount of security are available. Be sure to check into them.

Entries

Door sidelights have become popular as a way to make entries lighter and more attractive. These narrow vertical panels of glass on one or both sides of the door are often decorative, intended to give the entry a touch of luxury. They also provide an easy way for burglars to break in and unlock the door. (Security is discussed in more detail in Chapter 11.)

Unless planned carefully, sidelights can also cause problems with the light switch location—see the discussion in Chapter 5. The entry door is always in a load-bearing wall and framers tend to fill the area around the door and the sidelights with studs, leaving no place for the switches. This should all be figured out ahead of time.

Another sometimes-heard objection to entry door sidelights is that someone can see into the house. The windows are so small that putting an attractive blind or curtain over them is difficult but, without something, you lose your privacy at night. To let light in during the day while avoiding this fishbowl aspect of door sidelights, some builders use glass bricks or acrylic blocks instead of window panes. You may want to give some thought to the advantages of using other than transparent sidelights.

When You Have a Choice

A door that opens into a kitchen nook needs special attention. There is usually little room for a swinging door so builders will swing the door outside—which, as noted above, needs special care if it is not to be a security problem. Further, this arrangement leaves no place for a screen door.

The lesser of the evils is a sliding glass door that, even with its heat loss, is the more user-friendly answer. Another solution could be to have the outside door open into the family room adjacent to the nook. The disadvantage to this approach is that the outside door then opens directly into a carpeted area—which isn't good for carpets.

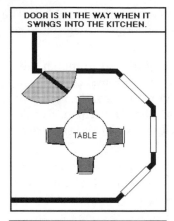
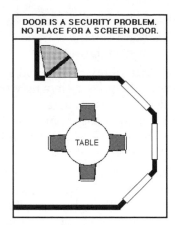

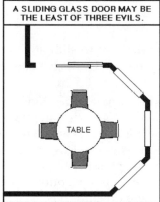

Doors opening into kitchen nooks need careful consideration.

If the nook can be made big enough that an in-swinging door doesn't get in the way, the problem is solved. Otherwise there will be an inevitable compromise, usually ending up with a slider.

Interior Doors

Since insulation value and security are of no relevance, interior doors are chosen for their appearance and it is generally desirable that they be compatible with the cabinets—a consideration when you look at the patterns and finishes of doors and cabinets.

Interior doors are usually hollow core. In a hollow-core door a frame of wood goes around the outside edge of the door and across the center. This is covered by a veneer of wood, molded hardboard, or vinyl. The center piece is used to mount the door latch. The rest of the interior of the door has a honeycomb of material to support the veneer on the outside.

Interior doors may be painted or finished to match the woodwork in the house. They may be a raised panel design or a single solid surface. The single surface doors are usually a hardwood veneer intended for staining and lacquer or varnish. Raised panel doors come in several different styles in wood and simulated wood hardboard or vinyl surfaces.

Hardboard raised panel doors can be painted or finished to look stained. White hardboard with white painted molding is a common finish although some builders choose a mixture of white doors and natural woodwork or vice-versa.

Bifold Doors

Bifold doors are often used in preference to sliding or hinged doors, particularly for closets. They have advantages and disadvantages.

Even when open, bifolds always block part of the doorway and can be a real annoyance when installed in closets that aren't walk-in. A disturbing characteristic of double bifold doors is that when you close one side the other side often pops open. This is, at least in part, because as you close one door, air pressure builds up in the closet that forces the other door to open. Louvered bifolds can alleviate this annoyance but don't put them on laundries.

Some designs in starter homes have the laundry room next to the kitchen with bifold doors to close off the laundry. You'll have a serious noise problem unless sound insulation is used around the laundry and the doors close snugly. Louvered doors cannot shut out the sound.

There are right and wrong ways to have mirrors on bifolds. The right way is to get doors which are basically mirrors with metal frames. The wrong way is to put mirrors on wooden doors because the hardware is not designed for this heavy combination and it may not last.

Pocket Doors

Pocket doors need different considerations. The pocket must be deep enough for the door to be pushed all the way in. Nothing else can go into the wall where the pocket is—no wiring, outlets, switches, or plumbing. Sometimes the pocket door may not be possible even when it otherwise makes the most sense.

Any place where a hinged door would be in the way when left open should be considered for a pocket door. Walk-in closets are good examples. You want a door you can close, but you don't want one you have to close from the inside to get to your clothes. (Hinged doors are obviously okay when they're not in the way when open.) The door into the toilet/bath part of a master bedroom suite is another place where a pocket door may be better. Still another is a walk-in pantry.

Although hinged doors are frequently used between dining rooms and kitchens, they shouldn't be because, when open, they take valuable wall space if swung into the dining room or they block counters if swung into the kitchen. They can also interfere with refrigerators. A pocket door, or even no door at all, is preferable.

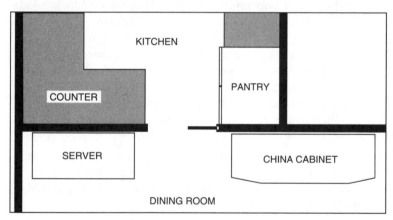

A pocket door (or no door at all) is the user-friendly design between kitchen and dining room.

Look at your floor plan carefully to see where pocket doors, rather than hinged doors, should be used . Replacing a hinged door with a pocket door on a floor plan may not be easy when the walls beside the hinged door mount the room light switch, an outlet, or contain water pipes (like to a shower head). It may take some rearranging to use the pocket door, such as putting the shower head at the other end of the shower, moving the light switch to the other side of the doorway, etc. If you can eliminate an inappropriate hinged door, however, it will be worth it.

Pocket doors are not always workers favorites. Like the carpenter who mounted the paper holder in the bathroom but had forgotten that there was going to be a pocket door in the wall.

And there was the building designer who got a call from the electrician wiring a house to say that there was a pocket door in the wall where the designer had an outlet. Caveat emptor!

Doorstops

The one place with any strength inside a hollow-core door is across the door center. Inside the door at that point is a block of wood (the rail) for mounting the door latch. When a door is mounted with three hinges, and the center one is used for a hinge-mounted doorstop, the stop hits the door over this block. When the stop is placed on the top or bottom hinge where there is no block inside the door, the stop can make a hole in the thin veneer.

When visiting in an area where interior hollow-core doors are mounted by two hinges, not three, my hosts showed me what happened within a week after they moved into their new house. A gust of wind caught the guest closet door and the doorstop punched a hole through the thin veneer of the door. Someone should have been more careful. But who?

Not the home user. The builder screwed up. Our house has the same kind of doors and doorstops. But we'll never have this problem because our doors are all mounted with three hinges, not two, and the stops are on the center hinges.

Often, this kind of doorstop is put on upside down. When it's installed correctly, the stop hits the door and the door trim squarely but, when it is wrong, it hits at an angle.

It is good practice to use hinge-mounted doorstops only where there is no alternative because they are much more likely to cause damage to hinges.

Doorstops that screw into moldings won't work when they are very close to the floor because the door can slide over the top of the stops. This can happen when moldings are not high enough to let the doorstops be mounted higher.

Double Doors

Designers and builders use double doors as entries into houses, master bedrooms, and dens to give a touch of luxury to a floor plan. After all, if your house is big enough to use double doors then it's obviously in the luxury class. Well, maybe.

The drawing shows one unfortunate master bedroom in a custom house that was proudly on display by the designer and the owner. One door opens across another, the light switch for the room is behind the door that normally opens, the closet door should be a pocket door so it doesn't swing out into the room or the closet, and there is a sidle toilet.

The location of light switches is a particular problem with double doors. When you have to take more than one step to reach a light switch after entering the room through a double door, you need a better arrangement. And, as discussed in Chapter 5, light switches should never be behind a door.

The light switch is most accessible when it's on the wall just past the end of the door normally used for going in and out. Therefore, be sure the finish carpenter knows which door will normally be used as the entry and which will be fixed.

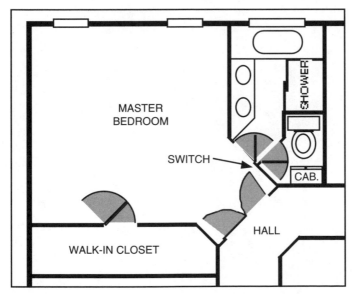

This disastrous MBR design happened when a poorly advised buyer hired an incompetent designer.

Doors and Furniture

Double doors, when opened, sometimes occupy wall space where you intended to place furniture.

Our Oregon home had double doors between the master bedroom and the vanity area. The designer's floor plan showed these doors opening into the vanity area—one door blocking the doorway into the toilet/shower room and the other door blocking light switches. The carpenter properly decided that the doors should swing into the bedroom—where they blocked two walls from having furniture on them. After living in the house for several months I removed both doors, leaving a totally acceptable archway and a much better furniture arrangement.

The problem with hinged doors interfering with furniture placement can also occur with single doors. Sometimes the solution may be a pocket door, as mentioned earlier; in other cases simply hinging the door from the other side may solve the problem. At times you may be forced to accept a compromise in door location. Take a good look at the floor plan to prevent door problems before your house is built.

Chapter 10

Windows

Energy saving is an integral part of the design and construction of today's houses. And a very important part of that energy savings has to do with the number and types of windows that are used. In this chapter we discuss the relationship between energy and windows as well as form and function.

I've tried to present the somewhat technical information in a user-friendly way. In the rest of the book, I've stayed away from anything that isn't self-evident, that isn't simple common sense. But in our energy-conserving environment, I felt it wouldn't be right to ignore the driving force that determines the window options available for today's homes.

Windows and Aesthetics

You know the real estate agent's expression "curb appeal." It's the impression prospective buyers have when they are taken to look at a house that's for sale. Windows make up a big part of that impression. They should fit in with the overall style of the house and they should make the house look inviting—a place that gives you a feeling of being lived in. And, it's different strokes for different folks—some like the large and multitudinous windows found in many of today's designs while others are happier with the conservative approach with more modest sizes and numbers of windows.

Inside the house, windows, probably more than any other feature, control the ambiance of a room. Correct windows are vital parts of how you will enjoy living in the house and, later on, of the resale value of the house.

Double-pane (a.k.a. double-glazed) windows are made with and without a decorative grid between the panes. This grid makes the windows appear to be made of a number of pieces of glass when, in fact, they are not. More and more of these windows use vinyl-clad frames for its better insulation. The vinyl is usually white but colors are available. (Anything other than white can limit your choice of house colors.)

Enough of aesthetics and ambiance. These are correctly left to architects and designers and are not the thrust of this book. But there is, as always, some give and take between form and function, between how windows look in a house and their costs in terms of energy and dollars.

Windows and Safety

Small windows high up on bedroom walls were common in houses many years ago but are not found today. Stories of people being trapped in their bedrooms in a fire, because the windows were too small or too high, caused codes to be changed. And even if not forbidden by building codes, common sense would dictate a safer design.

Today, bedroom windows must be large enough and low enough to get out of in an emergency. And there must be such a window in every bedroom in the house. (Codes spell all of this out in inches and millimeters.)

You might think that building codes would change little from year to year. Not so, particularly in the energy area where continuing improvements in materials make greater energy savings possible. One of those areas is windows. And well it should be because, as noted in Appendix B, windows represent the biggest single source of heat loss in most home designs today.

Window Construction

Windows are made of glass which is made of silica, the same thing that sand is made of—right? Well, yes. But there's a lot more to today's windows than that. The use of single pane windows is disappearing from new homes. (The building industry calls windows single or double glazed rather than single or double pane. We'll stick to the vernacular.)

Double-pane windows have these five things going for them:

- They make a house more comfortable.
- They don't weep or sweat (at least they shouldn't).
- They save energy in heating and cooling.
- Saving energy saves money.
- They make the house more saleable.

The downside is that they are more expensive initially.

There are three mechanisms by which heat goes through a window:

1) Radiation. This is heat your can "feel," like the sun's rays or when you hold your hands in front of a burning fireplace. This heat goes through the window like light does.

2) Conduction. This is the heat a solid material carries from one place to another, like from one side of a window pane to the other.

3) Convection. This the heat that is carried by air (or any gas) from one place to another. Heat is carried by convection between the panes of glass in a double pane window.

Modern window construction tackles all three of these to keep heat loss down.

The panes may be *low-e* (low-emissivity) glass. This involves a special manufacturing process that reduces the ability of the glass to pass certain heat rays (as well as ultraviolet that we'll discuss later) while still letting us see through the window. The idea of low-e glass is to let the direct sun's rays through for heating help in winter while stopping other heat rays so that the warmth generated in the house can't get out. In summer you have to depend on trees, drapes, shades, and roof overhangs to keep out the direct rays from the sun.

Conduction through the glass itself is a fact of life, but the materials around the edges of the window also conduct heat. This source of heat loss is tackled by using materials to mount the glass that are not good conductors—vinyl, for example. Wood and aluminum are also used. (Note how aesthetics get into the act here, too.)

The transfer of heat by convection between the panes of glass is inhibited by filling the space with a gas that doesn't do a good job at convection. Argon is the most commonly used but you may find other more exotic gasses. (A vacuum would stop convection entirely but it's not a practical solution.)

Putting argon between the panes of glass poses another problem—how to keep it there. In truth you can't, at least not with cost-effective manufacturing techniques. In time it will leak out, being replaced with plain old air. When this happens two things occur:

• The window will be a poorer insulator and

• Moisture will appear between the panes.

At that point you'll have three choices: (1) if the window is covered by a lifetime warranty, get the manufacturer to replace it, (2) replace it at your own expense, or (3) live with it.

Windows and Energy Codes

Windows are rated by a standardized process resulting in a U rating. This number is always a decimal fraction smaller than 1. The smaller the number the better the window in terms of energy conservation. Good windows will have a U-rating of less than 0.5. Better ones get down to 0.4 with manufacturers striving to get to 0.3. But note: a U-rating of 0.3 (very expensive) is not even twice as good as a rating of 0.5 (low cost). To spend a lot of money for a specially low-rated window is probably not effective in terms of either comfort (you'll never notice the difference) or energy savings (it'll take a lifetime to recoup the additional cost).

You may not have the choice, however, depending on where you live. In states where the idea of energy savings has a bureaucratic stronghold, there is continual pressure to reduce energy usage— even when it is not cost-effective. If you live in one of these areas, as I do, you may find low-e, low U-rated windows are required to meet energy code requirements. Because in most houses, windows account for more heat transfer between inside and outside than walls, floors, ceilings, or doors, some codes restrict the amount of glass, including skylights, you can put in your house. If you have these kinds of restrictions, some floor plans cannot be used.

On the other hand, these same codes may take into account that winter sun can be used to help heat the house and give "credit" for south-facing windows to reduce the amount of insulation you must have.

But don't be dismayed even if they cost more, because these windows make a house a whole lot more comfortable than the ones you were brought up in.

Windows and Fading

In southern California our house had a large west facing living room with the window glass in two tiers. We put drapes on the lower tier but, because of the irregular shape, had difficulty getting coverings for the upper ones. We did add a plastic coating which was supposed to cut down both the sun's heat rays and the ultraviolet—the part that does the fading. We thought it did a good job until we moved and found the original color of the carpet every place where furniture had covered the floor. The rest of the carpet where the sun could get to it was several shades lighter.

Today's low-e glass may be better than the plastic sheet but probably not much. It cuts out only a little over half of the ultraviolet. So, by all means, plan on trees, drapes, blinds, or whatever you can scrape up to keep the sun from beating into your house—unless, of course, you don't mind multi-shaded furniture and carpet. And feel good about it—you're not only saving your furniture and carpet, you're lowering the heating and cooling bill too.

Windows and Decorating

When windows are selected for their appearance, whether from the inside or the outside of the house, don't forget energy loss and fading—to say nothing about privacy.

Large windows will lose more energy than smaller ones with a resulting increase in heating and/or cooling bills. The more of the

Because they are hard to decorate, windows high on walls can be serious sun problems.

These upper windows will also be hard to decorate.

window you can cover with drapes when the outside temperature is too hot or too cold, the more comfortable the house and the lower your energy bills will be.

It is hard to get coverings for the upper parts of the two story high windows found in many modern home designs. And, of course, ultraviolet rays don't care if they come in through a round, trapezoidal, or triangular window above a more conventional lower one—they'll cause fading in any case.

Skylights

Most of today's house designs include skylights. They are used where windows would be inconvenient (as in small bathrooms), where windows aren't possible (as in an interior room that has no outside wall), or where they let in light while still maintaining privacy. There are penalties to pay when skylights are used. As windows, they are poor insulators. In general, it's not possible to shade them from the hot afternoon sun. When this is a problem, skylights will need to be made from tinted or low-e glass and/or they'll need some kind of interior blinds.

Where skylights are used in rooms like bathrooms that do not have vaulted ceilings, there is a shaft or "well" that brings the light down from the roof-mounted skylight. This well is often bigger at the bottom than the top thus spreading the light over a larger area in the room.

Skylights range from one to four feet on a side. Think carefully before using the smaller sizes. Many people don't like their penny-scrimping appearance.

When choosing skylights for a well that can trap a lot of heat, get types that can be opened to let accumulated hot air escape during hot weather.

Light Funnels

Relatively new for homes are metal tubes that funnel light from the roof down into a room. A small glass or plastic dome is mounted on the roof with appropriately shaped reflectors to capture the sunlight. At the bottom end a diffuser spreads the light around in the room. They are designed to mount in between existing structural members making them attractive for additions to existing houses where the original plans should have had a skylight but did not.

The light that is delivered into the room is *not* the same as you get from a skylight but rather has a distinctly cold metallic cast—which is not surprising since it depends on the reflection off the aluminum that makes up the unit. For new construction the tubes are less expensive than skylights with their wells. But, because of the cold, nonconventional color of the light, they will detract from the saleability of the house for many buyers. They are probably best left to the remodelers.

Window Woodwrap

In some parts of the country, window openings are usually encased ("wrapped") in wood while in others builders offer it as an option. In Arizona, California and other areas with southwestern architecture, the usual approach is to provide a wooden sill or even no sill at all; the opening is simply handled as a continuation of the wallboard. These regional differences can be explained by the weather differences in combination with older single-pane windows. In climates with damper, colder winters old windows wept or sweat and wallboard or plaster was unsuitable around windows.

Where today's codes require double-pane windows that don't sweat (or at least not much), the functional need for woodwrap has been eliminated everywhere. Aesthetically, woodwrap has a richer, more finished appearance than does wallboard around a window and this has helped to keep its popularity in areas with more traditional architecture.

Woodwrap is an expensive proposition. For example, a tract builder in Tacoma, Washington offered this amenity as an option with a $2000 price tag on it! You might give some thought to whether you really need it, particularly when you plan to paint the wood white against a white wall and then cover it with a valance and drapes so that only the sill shows.

Chapter 11

Security Considerations

While we may bemoan the circumstances and societal failures that are responsible, residential burglary is a fact of life. This chapter discusses some of the things you can do to cut down on the susceptibility of your home to break-ins. The most obvious is an expensive alarm system that includes sensors that trigger whenever there's an unauthorized entry. But there are other small, inexpensive things you or the builder can and should do that will help. These are appropriate even with a central alarm system.

Securing Doors

Deadbolts are put into new houses routinely. They operate by moving a heavy piece of metal (the bolt) from the door through a hole in a metal striker plate fastened to the inside of the door frame. If this plate is not well secured, a hard shove on the door will break the plate loose and the whole door swings wide open. Unless instructed to do otherwise, the carpenter who installs the striker plates will usually use short screws that extend only into the door frame. He should use much longer screws (2½-inch minimum) to get all the way into the wall studs behind the frame.

Door hinges should also be mounted with screws that reach into the studs behind the door frame. This isn't usually done. Both the striker plate and the hinge screws will be stronger if the gaps between the frame and the studs are snugly shimmed. To beef up the door installation even more, horizontal blocking should be used between studs on either side of the door. These recommendations are courtesy of the Portland, Oregon Police Department.

You should insist on having doors put in right. It doesn't cost much at the time of installation but it virtually eliminates one method of easy access for burglars.

You can and should put in the longer screws if they are not put in the house initially but, once the wallboard is in place, you cannot do the shimming or blocking that makes them stronger. It's better if the installation is done properly.

Glass In and Around the Entry Door

The utilitarian purpose of door sidelights is to provide illumination in the entry hall or foyer during the daylight hours. Decorators, suppliers, and builders have seized on this need to make them a decorative and attractive part of the entry way. Unfortunately they ignore how easy they have made it for people to break into our houses.

Beware of glass either in the door itself or in the sidelights beside the door. These features make it possible for the burglar to

Sidelights are attractive. They also make it easy to break in.

break the glass and reach in and open the door from the inside. You could stop this by using deadbolts that require a key to open from the inside. However, this is dangerous if there's a fire and someone is inside the building. (And it's not permitted by building codes.) The safest approach is to not put glass where it'll let someone get in the house easily. And, for the sidelights, this can usually be accomplished without significantly degrading the appearance of the entry.

The problem doesn't exist for double doors because the deadbolt can't be reached from any sidelight location. For single entry doors, one of two things can be done—and this should happen at the house design stage. For a single sidelight, either put it on the hinge side of the door or space it far enough away that the deadbolt cannot be reached from the sidelight. For two sidelights, space both of them, or at least the one on the deadbolt side of the door, away from the deadbolt. Note that this approach does something else, too; it leaves a place close to the door for the entry light switches—they don't have to be several feet away on the other side of the sidelights as you'll see in many of today's houses and as is discussed in Chapter 5.

And don't forget the door itself. Many doors include decorative glass inserts; some decorators feel that the more ostentatious the house the fancier the front door needs to be. When these inserts are within easy reach of the deadbolt, the situation is akin to that of close-in sidelights: the door can easily be opened once the glass is broken. When the glass is higher up on the door it's no longer a means of easy entry.

Garage Sidedoor

When there is a garage sidedoor, the door hinges and the deadbolt striker plate should be mounted with long screws as discussed earlier.

Garage sidedoors with windows are security concerns for two reasons: they let people look into the garage from outside and they can be broken to get at the locks on the inside of the door. Since the sidedoor is almost never on the street side of the house, it is a more attractive target than the sidelights on the entry door, for example. For these reasons, it is strongly suggested that the sidedoor not have glass. If you feel you really need the light such a window lets in, then use a heavy metal mesh to prevent someone from reaching their hand through the door after the glass is broken.

The same considerations apply to any windows in the garage that are large enough for someone to crawl through.

Garage Door Openers

When you are going to be gone for an extended period, it is a good idea to make your garage doors inoperative. Four ways for doing this are:

• Have a separate wall switch in the garage that turns the power on and off to the openers. This is the most convenient way to control the doors. It needs to be done when the house is wired.

• Have the electrician put the door openers on their own circuit in the circuit-breaker box. Then turn off the breaker to disable the doors. (This assumes the circuit-breaker box is inaccessible from outside the house.)

- Get a cord with a switch in it and put this between the outlet and the power to the openers. Let the cord hang down so the switch is convenient.
- Unplug the power to the openers when you are gone. This is the least convenient, particularly when it requires a ladder to reach the plugs.

Between the House and the Garage

In some areas (the Puget Sound is one), it is common to not include a deadbolt or any other kind of lock on the door between the garage and the house. While it is true that once someone gets into your garage it is easier to break into the house, a lock on this door will make it more difficult and will make the house more secure. If you find a house that doesn't have a lock on this door, then it's suggested that the builder add one, preferably a deadbolt; either that or plan on doing it yourself once you move in.

Fire Safety

Smoke detectors are a code requirement for safety reasons. These are usually installed in the bedroom wing of the house because of the concern of fires trapping people in their beds. Don't hesitate to install extra detectors in rooms with tall vaulted ceilings. The tendency of heat to rise can make these detectors more effective in some cases than those in bedroom halls.

Some jurisdictions (Livermore, California for example) have a code requirement for automatic fire sprinklers in two-story houses. Others have special requirements for houses abutting open areas where grass or other wild fires can be a hazard.

Chapter 10 notes that there are code requirements on bedroom windows to make it possible to escape in case of fire.

Security Lights

When the house is being built, it's relatively easy to have wiring installed for security lights around your home. Where there are no street lights, it's convenient for guests at night if there is a light with an infrared sensor in front of the house. It's also a way of dissuading unwanted visitors.

You can have such lights installed by the builder or simply have the wiring put in place so you can mount the lights and sensors at your convenience. A switch is needed someplace, usually in the garage, that can disconnect these lights. As an alternative, have them on their own electrical circuit so that they can be turned off from the circuit-breaker box.

An Alarm System

While these and other techniques can be used to make your house more secure, the ultimate solution is a security system. You don't have to have this installed when the house is built, but putting the wiring in place before the wallboard goes up will make it less expensive to add the rest of the system later. If the wiring is installed during construction, it's possible to hide the wires. If you want this done, you will need to discuss it with the builder and with a representative of an alarm company.

Alternately, there are wireless systems that can be installed at any time. They are not as reliable as the wired systems since each sensor is more subject to circuit failure as well as battery failure. It is, however, a way out if you decide you need an alarm system at some time in the future.

Chapter 12

Other Whole-House Items

A number of critical decisions you will have to make involve the choices of materials. Some general guidelines start this chapter. These are followed by discussions of interior items including finishes, wallboard, baseboard moldings, floors, ceilings, and sound insulation.

Materials

Suppliers of materials for new homes are in a highly competitive environment and are always using new ideas and technology to come up with something that is different, attractive and saleable.

As the consumer, it's hard to know when it's appropriate to get the latest thing and when it's prudent to stick to the traditional. Home building is a very conservative business for the reasons noted in the *Introduction*. Change comes slowly because no one wants to or can afford to be the guinea pig. Yet, at the same time, better products are continually coming into the market—products that could make your home a better, less expensive place to live.

In the area of energy savings, some of the decisions have been taken out of your hands by changes in building codes. You must insulate your walls and use high-tech windows to save energy. And if you are building where there are earthquakes you'll have another set of criteria that must be met. But oftentimes you'll have to make decisions where what you really need is the proverbial crystal ball.

If the new product is supposed to save money, balance this against the long-term cost if it proves to have a limited life. Some examples that come to mind:

- A composite house siding made by Lousiana Pacific was supposed to be the answer to the dwindling supply and increasing costs of traditional siding materials. Yet in a few years of actual usage the siding turned out to not last. The result was a significant loss to Lousiana Pacific and to we home owners who happened to have had it on our homes.

- Some twenty years ago a cultured marble countertop with integral basins was a way to get a low-cost, handsome bathroom vanity. But after a few years the surface of the basins developed tiny cracks caused by the repetitive temperature changes of hot and cold water. In this case there was no Lousiana Pacific to fall back on and the home owners bore the cost of replacing the counters and sinks. (There is now better quality control on these items which the industry hopes has solved the problem, but *see* Chapter 20 for other solutions.)

- Polybutylene water pipe for inside homes offers a number of advantages as discussed in Chapter 6. But it wasn't always so. The connectors used when PB was first introduced didn't hold up with time to the dismay of the manufacturer and home owners. Again, it took time to find the problem and fix it.

Steel framing to replace the traditional wood construction has some obvious advantages in terms of stability—no warping, shrinking, termites, rotting, or fire. It's installed price is about the same. What most of us don't know are the potential disadvantages nor are we likely to find out from the suppliers and builders of traditional materials.

Low-emissivity glass for windows is now an energy code requirement in many areas (*see* Chapter 10). But at what point are further improvements no longer cost effective? And to whom do we turn to find out?

So, unless your crystal ball is working well, you'll find yourself in the position of making decisions about products that you know little about and what you can find out may be only part of the facts. Here are a couple of guidelines:

- Try to find actual experiences people have had in their homes. If it's a new product, find a reputable supplier who has competing products and *try* to get an honest appraisal of the situation. (The quartz resin composite sink mentioned in Chapter 13 is such a product. It appears to be an excellent option, but still lacks general acceptance.)

- If the new product is supposed to save money, get some actual quotes for the new one versus an old one and see if there is actually that much difference in the installed cost. To say a product is half the cost but to not include the additional fabrication costs associated with it isn't the whole story. A competitive bid will give a much better price comparison. Then you can make a better assessment of initial cost saving versus the long-term risk. (The solid surface veneer countertop discussed in Chapter 13 is an example. It's new and relatively untried. The basic material is less expensive but its fabrication is more difficult.)

Finishes

In sharp contrast to houses constructed even fifteen years ago, the trend today is to lighter colors and woodwork in homes although the richness of darker cabinets is favored in some regions. It's a matter of aesthetics and it's all up to you.

Local tradition and custom are strong factors that apply to decorative wood in the house: baseboard moldings, crown moldings, wainscot, doors, and windows.

Today there is an increasing trend toward painting the woodwork, moldings, woodwrap, and even cabinets and doors. It is the fashionable thing to do. Traditionally, natural wood has been associated with luxury; you painted only when you couldn't afford wood that could be left natural. Or you painted an older house when it was no longer feasible to take care of mars and cracks any other way.

White or near white paint does make a room feel brighter and more open. This trend to white also extends to countertops. It's the safe thing for builders to do because almost nobody objects—we've become accustomed to see white everywhere in homes. But too much white may remind you of a doctor's office, a hospital, or a service station washroom. A popular compromise is to stain the woodwork a light shade and paint the walls white.

The decision whether to paint or to stain and lacquer is very much one of personal preference.

Walls

Gypsum wallboard is the most commonly used interior wall material. The sheets are nailed, glued, and/or screwed to wall studs and ceiling joists. There are gaps where edges of the board meet and there are other irregularities at corners. These are all covered by a thin paper tape and a plaster-like "mud" and made as smooth as the craftsman deems necessary.

Corners may be square or rounded. Generally rounded corners will be a little more costly, but are preferred by many people because of the softer feel they give to a house. It's a matter of personal taste and budget.

Wallpaper

If you plan on having a wall papered, two things are important:
1. Don't let the wallboard people texture the wall. Be sure that someone marks the wall in big red letters to keep this from happening.
2. The corner where the wall meets the ceiling should be straight and even. If the corner is uneven, it is virtually impossible for the wallpaper installer to make the paper look straight across the top of the wall.

Finishing the Wallboard

In some areas wallboard is simply smoothed and painted and in others it is finished by texturing. This is the process of putting mud on the wall in an uneven pattern. A common finish is called orange peel because of the look of the texture. Ceilings generally have a deeper pattern. Your builder can show you different wallboard textures or you can shop around for the ones you like best.

Building codes require that walls be sealed to make them vapor proof. There are several ways to do this but one is to use a special paint. If you and your builder choose this technique, you'll need to choose your wall colors from the paints that have this property.

Make sure that the board used on ceilings is designed for ceilings; wallboard should not be used. Ceiling board is stiffer and won't sag between joists.

Baseboard Molding

No Baseboard at All—Builders, when they build houses for someone else, are constrained by their agreement with the buyer or, if it's a spec house, by their estimate of what will sell best. When they build a house for themselves, they are freer to do what they like best. One thing is to eliminate baseboard molding in the carpeted areas of the house. This requires that the wallboard be finished all the way to the floor. It also requires the use of hinge-mounted door-stops because there's no molding into which to screw the normal doorstop. And this, in turn, also means that hollow-core doors must be mounted with three hinges, as is explained in Chapter 9.

The carpet covers the area where the wallboard meets the floor, creating a clean, crisp appearance. It's certainly a better looking finish than those seen in most tract houses which look as if the builders picked up old boards from the scrap heap to use for baseboard moldings. Eliminating baseboard molding also eliminates the problem of getting a good finish on it.

Wooden Baseboard Molding—There's no reason you shouldn't have a decorative molding of reasonable size or to not have any at all.

- In some areas the baseboard moldings in tract houses are just plain $\frac{1}{2} \times 3$ inch boards, sometimes with a 45-degree bevel along the top edge and sometimes not. They are painted, not stained. They look cheap and detract significantly from the appearance of the house.

- In other tracts, builders put in baseboard moldings that are too small, some no bigger than $\frac{3}{8} \times 1\frac{1}{2}$ inches. More than one model was seen where the door slid over the top of a doorstop because there was no place to mount the doorstop high enough to catch the door.

Rubber Mop Boards—In some tracts builders have reverted to an earlier practice of putting rubber mop boards in kitchens, bathrooms, and laundries. Many years ago when a mop consisted of many sloppy twisted pieces of rope-like cotton, rubber mop boards were necessary to protect the bases of cupboards from excess mop water. This gave rise to the use of the 3- to 4-inch rubber mop boards. Today's mops don't do this and wooden baseboards with a reasonable finish are fine. However, if you feel more comfortable with the utilitarian rubber mop boards, they are still available.

Rounded Wallboard Corners—Rounded wallboard corners pose a problem for the finish carpenter. The molding should follow smoothly around the corner, but this obviously won't happen. The usual solution is to cut a small piece of molding with 45-degree mitered ends and glue it in place across the corners. Gaps will still be noticeable in unpainted molding but if the molding is to be painted, the gaps can be caulked resulting in a more finished appearance.

Small pieces of wood carved to match the baseboard and curved to go around rounded wallboard corners are available, at least for some types of molding. If matched to the molding reasonably well, they are more attractive than a straight piece at 45 degrees.

With rounded corners and painted moldings, there is still another option; the moldings are simply mitered and cut so that they meet squarely at the corner. This leaves a space between the molding and the rounded wallboard corner. This is filled and painted over. The advantage is the lower cost for the finish carpenter work. The approach is useful only with painted molding although some low-end builders will do the same thing with wood finish moldings by filling the gap and painting it to try to match the molding.

Vinyl moldings are available in wood finishes and are used in some low-end houses. If you are trying to save every penny, you may want to take a look at them.

Gaps—If you can, check the workmanship after framing and again after wallboard installation to see that walls are truly straight at the bottom where the molding goes. Where they are not, you'll have small but noticeable gaps between the molding and the wallboard. If there are problems, the builder should have them fixed before the finish carpenter starts work on the baseboards.

If the molding is stained and lacquered, gaps along the baseboard will be clearly visible. They can be fixed but it may involve the wallboard mudder, the finish carpenter, and the painter. And, if the builder has to pay for it, he'll no doubt argue with you about whether it's necessary.

When the baseboard molding is painted, you have a much better chance of getting a finish job that doesn't draw attention to itself. The finish carpenter or painter can caulk the gaps and paint over them. The imperfections will still be there but will be covered up.

Closets

If building designers spent as many hours doing housework in the houses they design as they spend in designing them, there are some things they'd be quick to change. The amount of closet space and closet design are two of them.

Walk-in Closets

Hinged doors are the most commonly used for walk-in closets but these should not open into a closet in a way that the door blocks access to any of the closet's interior. Apparently this is easier to say than do considering the large number of closets built this way. One house in six was this way in the review of almost seven hundred floor plans.

Closets and Accessibility

Many closets that use sliding, hinged, or bifold doors are designed so that they are partly inaccessible! It's probably not deliberate but it's not user-friendly either. The problem is that the front openings of the closets are too narrow. It's not unusual to see closets with over a foot of closet past the end of the door opening. Add 6 or 7 inches of an open bifold door in the front of the opening and you can have close to 18 inches of inaccessible closet on at least one end. Try getting a coat or dress out of the end of that closet and you'll be more than a little annoyed.

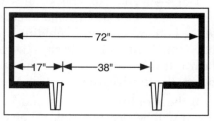

The problem arises because doors come in preset widths and these may not match the space the designer has available for the closet.

When doors are not as wide as the closet, they severely restrict access.

Better planning should make it possible to have closets whose sizes match available doors. When looking at models, keep an eye open for these ill-designed closets. You'll be surprised at how many of them there are—it occurred in 25 percent of the houses in the floor plan review.

Closet Interiors

It used to be that a closet had just a shelf and a pole for hanging clothes. Some still do but other, more useful, arrangements are seen today. These include built-in shelves for shoes or whatever else the home user may want. Poles may be two levels in height for greater utility.

Some builders use stiff vinyl-coated steel wire instead of shelves in closets. Again, these may or may not appeal to you. But not in pantries, please (*see* Chapter 18).

Here's a little problem, easy to fix when it arises, but annoying when it happens. In less than two years in our new home we had clothes dumped on the floor seven times by breaking pole sockets. Pole sockets are the plastic, wood, or metal pieces that hold the ends of the clothes rods or poles. Ours were a translucent white plastic that, over a short period of time, de-plasti-cized, lost their strength and split in two. If you find these in your house, keep your fingers crossed. If the first one breaks, bite the bullet and replace them all because it'll be just a matter of time until the rest go. (We've gone over to wooden sockets that won't de-plasticize.) If you can take care of this ahead of time, the problem won't arise.

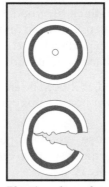

Plastic pole sockets don't last: before and after less than a year.

Floors

The common materials used for floors are carpet, linoleum, hardwood, ceramic tile, granite, marble, and field stone. With any of these it usually doesn't pay to buy the cheapest thing around. It won't last.

Subfloors

If you choose hardwood flooring that has to be nailed down, make sure it's nailed to plywood rather than particleboard. Particleboard won't hold nails and the flooring will loosen with usage.

Unless special precautions are taken, unyielding materials, like ceramic tile, marble, and other stone, will come loose or crack. These materials should be laid on a very solid base. Most subfloors don't meet this requirement. With post-and-beam construction and its 2-inch deck under the flooring, the deck lumber can be expected to dry out, warping in the process. Further, changes in temperature and humidity will cause the boards to move.

Floor joists made of sawn or dimensioned lumber are subject to the same drying and warping. Composite ("silent floor") joists are more stable over time but are still subject to changes in temperature and humidity.

To get the necessary rigidity, special treatment under ceramic tile or stone should be used. This may consist of a concrete-like foundation or mud, thick plywood, or backer board made of concrete mortar with fiberglass mesh reinforcing. Ceramic tile or stone should not be laid on particleboard or interior grade plywood that is not stable with changes in humidity.

If plywood is used, it should be exterior grade measuring at least ¾ inches thick. Backer boards are typically ⅜ or 7/16 inches thick and do a better job.

Slab floors, of course, won't shift but slabs can develop hairline cracks which cause tile or stone to crack if it is laid directly on the slab—which it usually is.

Tile

Entry—Ceramic or stone tile entry ways can cause accidents when someone with wet shoes comes into the house. The glaze on many tiles is smooth and gets slippery when wet. Tiles made for entries should be either unglazed or have a texture to reduce their slickness.

To test a tile's suitability for your entry way, take a pair of rubber-soled shoes, put some water on the tile, and test the traction. The people at the tile store should be able to advise you on the best purchase.

The reason for saying "should" is that we have slippery tile in our entry—purchased before I wrote the book, obviously.

In any case, if you are planning tile (ceramic or stone) in the entry, be sure to have a large absorbent rug at the door which will catch most of the water that may be tracked in. It may help prevent an accident.

Kitchen—A reader in western Kentucky relates that ceramic tile kitchen floors are the regional choice—linoleum is considered tacky. From a strictly functional viewpoint tile is a poor choice. It is potentially slippery, harder to maintain (scratches and grout stains), harder on the feet, and, because of its unevenness, a problem for furniture when used in the nook. Linoleum is better and modern wood floors are an alternative if you live in an area where linoleum is just unacceptable.

Baths—Tile in bathrooms doesn't have the same potential for problems as in the kitchen. Just be sure not to get one that's slippery when wet. Note that this means that pretty, shiny glazed tile is out. Marble also makes a beautiful floor, but don't have a shiny surface put on it or it, too, will be slippery.

Squeaky Floors

Squeaky floors are a source of irritation. Often there's little that can be done about them—not, at least, at a reasonable cost. And they are a real turn off for potential buyers at resale time.

Particleboard is usually used under carpeting. It expands and shrinks with temperature and humidity so, if it is laid without adequate space between pieces, squeaks can occur when pieces expand and rub together. This is something that is hard to write into specifications but if you don't see small spaces between boards when the house is being built, don't hesitate to speak up.

Squeaky floors can develop as the lumber in a house dries. Unless kiln dried, lumber shrinks or warps with time, leaving a gap between the beam or floor joist and the deck or subfloor just above it. When someone walks on it, the floor or subfloor will give and move up and down on its nails. The rubbing on the nails can cause squeaks.This doesn't usually happen until the house is a year old or more. Then you're on your own unless you have an agreement with the builder that he's responsible for more than the year of the normal move-in warranty. The best insurance is to take all the reasonable steps you can to eliminate the sources of the squeaks during construction.

You may not be as lucky as a neighbor who, after about nine months in his new house, couldn't walk in his bedroom or dining room without squeaking. In his case the problem was covered by his 1-year warranty. The carpet had to be pulled back and the subfloor re-nailed. A few months later and any squeak fixing would have been on the owner's nickel.

Not all floors will have squeaks. With slab floors, for example, there's nothing to squeak. There are different squeak mechanisms for post-and-beam construction than for floors that use joists.

With Post and Beam

This has been the most common crawl-space construction used in the western states. Typically four-inch wide beams are spaced on four-foot centers and supported every eight feet by posts that rest on concrete piers under the house. Two-inch lumber, typically 2"×8", is then laid across the beams to form the deck on which the house is built. Plywood, particleboard, or tile backer board is laid on the deck as the subfloor that holds the carpet, linoleum, or other flooring surface.

There are a number of places where squeaks can develop. An obvious one is where the deck is nailed to the beams. As the beams dry and warp, or as the deck lumber warps, nails will pull up and then squeak when someone walks on the floor above. Warping of the deck lumber can cause the subflooring nails to pull loose, another place for squeaks. Ideally, kiln-dried lumber will be used. If not, the best bet is to do the best possible job of fastening the deck pieces down, using screws or lots and lots of nails.

With Floor Joists

Some construction uses floor joists with much longer spans than are feasible with beams. Joists are used on second floors, in houses with basements, and in other places as an alternative to concrete piers and posts. Joists may be either sawn lumber or they may be manufactured joists made like I-beams in which multiple pieces of wood are glued together to form the joists. These are lighter, stronger, and longer than a single piece of sawn lumber. As the price of sawn lumber increases, these manufactured joists become more attractive economically. Joists are typically installed on 16-inch centers and plywood is laid across them as the subfloor. Gluing the subfloor to the joists eliminates one possible source of squeaks.

These manufactured joists are advertised as the way to have a silent floor. Since they're made with dried wood, as is the plywood on top of them, there's nothing to dry out and warp, hence nothing to cause squeaks. But there are other sources of squeaks with these floors. Joists are held in place with specially-shaped metal hangers. If a joist is not properly fastened, walking on the floor above can cause the joist to move in the hanger, causing a squeak.

Talk to your builder about the use of silent joists versus those made with sawn lumber. Or check to see what's under a house you may be considering. (Slab floors are no problem but squeaky floors

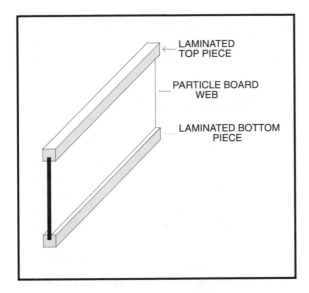

LAMINATED
TOP PIECE

PARTICLE BOARD
WEB

LAMINATED BOTTOM
PIECE

A 1.5" by 12'
'silent' floor joist.

can happen in any house that has an upstairs.) Where the span is long, the manufactured joists may be the only choice. If manufactured joists are not used, then consider having the subfloor fastened down with screws rather than nails.

The effects on the crawl space of using manufactured joints under a house are discussed in Chapter 23.

Ceilings

The living and dining room ceilings of many of today's houses are either two stories tall or they are vaulted These ceilings are all high and give a feeling of airiness. But watch out for ceilings in other rooms. Large rooms need high ceilings or they will seem oppressive.

- One house I saw is on a hillside lot with a nice view of the city. The great room has a ceiling that is 7 feet-9 inches high. It has a feeling of being even less than that. It is oppressive. The builder had trouble selling the house. That low ceiling in a big room was a real turnoff.

- On the other side of the coin is a house with a small living room about 10×12 feet but with a 10-foot ceiling. The high ceiling actually makes the room feel even smaller than it is. It pays to have the ceiling height in keeping with the size of the room.

Note that "8-foot" ceilings are frequently a few inches lower than this. Builders do this to save on studs that come in 8- and 10-foot' lengths. To be able to use the 8-foot piece, the inside of the room is made smaller.

Attic Storage

Crawling around the lumber in the trusses in our attic getting things out and putting them away every year at Christmas time reminds me again about the importance of having suitable attic storage space. If you plan on attic storage, this is certainly one place that, given the opportunity, you should be more than willing to spend a few dollars to have it done right.

For attic storage you should:

• See that the attic area is suitably stick framed or that special attic trusses are used to leave a maximum of open area.

• See that any attic insulation is roll or batting (not blown in).

• Spell out how much and what type of attic flooring is to be used. ¾-inch plywood or a dense particle board is fine.)

• Spell out the means of access you want, such as a door off of an upstairs hall or bedroom or a pull-down ladder in the garage. (Don't skimp on the size of the opening, you'll want to be able to get both a big box and yourself through it.)

• Use screw-in bare-light sockets mounted to a rafter—they are entirely adequate for attic lighting. The switch for the lights should be mounted at the entry to the storage area.

Quiet Please!

When we bought our new house, we didn't notice that the location for the entertainment center in the family room would be just a four-inch wall away from the head of the bed in the master bedroom. Going to bed early means using ear plugs.

Besides the obvious need to insulate bedrooms from adjacent family rooms, it's also a good idea to put sound insulation in the walls around bathrooms so that you or your guests are not awakened by early risers or middle-of-the-night users.

Laundry room walls are good places for sound insulation, especially if the laundry is in the living area where people are likely to be when the machines are operating.

Sound insulation is relatively inexpensive to install before wallboard is put in place. It's better to think ahead than to consider a retrofit. While it's possible to insulate the walls after construction, it is much better to have it done when the house is being built. The following techniques, used in apartments and zero-lot-line houses, can be used to significantly reduce the noise level inside a house.

Sound goes through walls in two different ways:

1. Through the space between studs. This is controlled by filling the space with fiberglass batting similar to that used for thermal insulation in exterior walls except there is no moisture barrier.

2. Through the wooden studs. There are three methods to reduce the sound carried this way. All three separate the wallboard from the studs and all result in thicker walls.

• **Double studs.** A staggered double row breaks the path through the studs. All spaces can be filled with fiberglass batting.

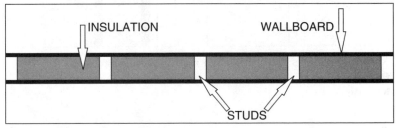

Simple wall insulation is a good noise barrier.

Double studs reduce the transmission of sound through them.

• **Resilient channels.** These are special U-shaped metal strips that are laid horizontally across the studs. The wallboard is screwed to these strips. Fiberglass batting is still needed for the space between studs. Using the channels on both sides of the wall obviously improves the performance.

• **Soundboard.** Soundboard goes between the studs and the wallboard, is relatively inexpensive and comes in 4×8-foot sheets. It
provides insulation for the whole wall including the studs. It,
too, can be used on both sides of the wall. Fiberglass batting will
provide additional sound deadening.

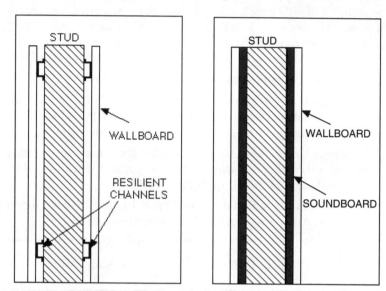

*Resilient channels and soundboard reduce the noise transmitted
through the studs. Soundboard also helps insulate the space
between studs.*

Part III

The Kitchen

The kitchen receives what may seem to be a disproportionate amount of attention in this book simply because it is the room about which far more decisions have to be made than any other.

Yet it is all rather simple. First there are the basic user-friendly rules. As with other parts of the house, they are all just common sense. You don't need to be a kitchen designer with years of experience to understand them. But do be careful—user-friendly design is not a major concern of architects, building designers, or kitchen designers as you can see when you look in magazines and books on kitchens, at design catalogs, and at existing houses.

It is the room in the home where more work is done than all the rest put together. The emphasis should be on user-friendliness—minimizing the amount of effort needed in the kitchen and making it a more enjoyable place to work and live in. Armed with this information, you can make the informed decisions that are right for your lifestyle.

Kitchens are areas of inevitable compromise. They must fit with the rest of the house and there are often conflicting objectives—aesthetic, as well as functional. A nice view, a special decor, or the desire for a large informal gathering area with the kitchen at the center—these are a few of the possible criteria that come into play.

In Part III you will find discussions of the various parts of the kitchen—appliances, cabinets, counters, etc.—in terms of some of their characteristics and good and not-so-good ways to use them. The last chapter in this part puts it all together with discussions about basic kitchen layouts, pointing out good features and potential problems. It's work, but that's what makes dreams come true.

Chapter 13

Sinks and Counters

In this chapter we look at the different materials that are available for kitchen countertops and how sinks can mount in them. Materials used for counters are laminates (popularly known by the trade name Formica), ceramic tile, solid surface (such as Corian), natural stone (usually granite), and solid surfacing veneer (a new material which is a hybrid between laminates and solid surface materials). There are functional, aesthetic, and cost distinctions among them.

Regional Differences

Regional perceptions about the different materials vary widely. In some areas ceramic tile is the most commonly used, laminates are considered "cheap," and solid surface materials are too expensive except, perhaps, in very high end models. In other parts of the country laminate is the material of choice for most people, with tile being saved for "show" places and solid surface for luxury homes. In still other areas it is common to see a mix, with tile on islands that have cooktops and laminate in the rest of the kitchen. Stone is often used instead of ceramic tile for cooktops and work areas.

Sinks

Popular sinks and how they are mounted vary by region. In some areas you'll find mostly cast iron drop-in or steel self-rimming sinks sitting on laminate counters while in others tile-in or undercounter mounting is used almost exclusively. Regardless of the regional differences, some sinks and their mountings make life just plain easier in a kitchen.

Mounting

There are three ways a sink can be mounted in a countertop: drop-in, tile-in, and undercounter. In every case the sink is supported at its edges.

Drop-in—These sinks are mounted by cutting a hole in the countertop and dropping the sink in from above. The edge or rim of the sink sits on top of the counter. This rim may be up to half an inch high in the case of cast iron or it may bethe thin rim of a stainless steel sink.

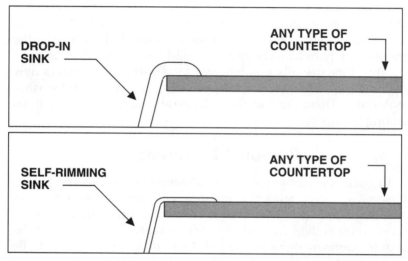

Counters with sink lips or rims on top of them are harder to keep clean.

Drop-in cast-iron sink on tile counter. Note the sink lip.

Because it sits on top of the counter, the rim is in the way when wiping or drying the countertop. Further, the joint where the lip or rim meets the countertop inevitably requires special attention to keep clean.

Drop-in is the most common type of mounting, not because it's the best or makes the least work in the kitchen, but because it's the only way that sinks can be mounted on laminate, the least expensive countertop material.

Oone counter manufacturer say told me that he has developed a means of mounting a sink under a laminate counter. I haven't seen one and remain skeptical—particularly about how well such a combination will withstand the test of time.

Drop-in sinks are also used on tile and solid surface counters even though more user-friendly mounting techniques could be used. This appears to be a matter of regional preference—and not to the advantage of the home user. It's as if builders and buyers got together and agreed that that's how they're going to do it and forget function.

Tile-in—This technique is useful only with tile counters—ceramic or stone. The sink is made with a special rim which is rectangular and not rounded at the edges or corners. The top surface of the sink is level with the tile making it a continuation of the countertop. The interface between the sink and counter is the line of grout around the edge of the sink.

The result is a countertop and sink combination which is easy to keep clean. The counter can be wiped directly into the sink without fighting the rim of a drop-in sink.

If you dislike ceramic tile counters in general, consider stone which can also be used with tile-in sinks but does not have the wide grout line. Either way, day-to-day work in the kitchen is less onerous than with a drop-in sink.

Undercounter—There are three ways a sink can be mounted under the counter, all involving how the sink and the countertop are joined together.

• With a tile counter, special quarter-round tiles are used at the edges where the sink and counter meet. This provides a smooth transition from the tile to the sink edge under it.

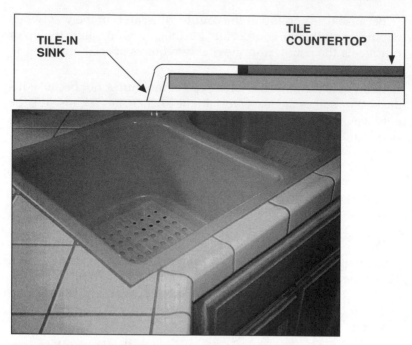

Tile-in sinks mount level with the top of a tile counter.

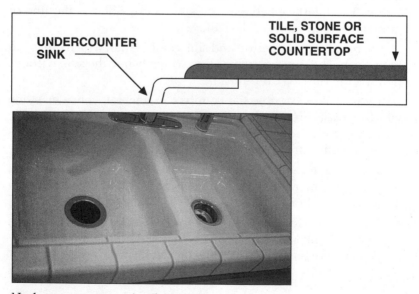

Undercounter mount in tile counter.

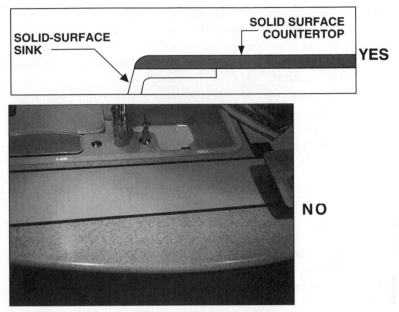

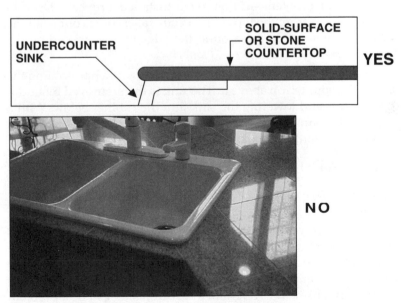

Sacrilege! A **drop-in** *sink on any solid-surface counter is absurd.*

Mon Dieu! A drop-in *sink on a granite counter. (Mount sinks* **under** *these counters.)*

- When solid-surface material is used in both the sink and the countertop, the interface between the two materials is at the underside of the counter where it meets the top of the sink. Properly made, there is no visible line and there is nothing to get in the way when cleaning.

- When a different sink material is used under a solid-surface or stone counter, another technique is used. The edge of the counter sticks out past the inner edge of the sink so that the sink is recessed slightly back under the countertop. Again there is nothing in the way when cleaning. (*See* diagram.)

I find it disheartening to walk into a model or show home and see a drop-in sink on a solid-surface counter. Whoever makes that decision does it from ignorance, not because of the cost—after all, when there's a solid-surface counter, expense cannot be a major consideration. And this is always done in upscale houses where you'd think that someone involved in the process would see how incongruous it is.

Sink Replacement

At some point sinks may have to be replaced and this should be no particular problem—if they were installed properly. Drop-in sinks come out from the top. A tile-in sink also comes out from the top with the line of grout around the edge of sink providing the breaking point between tile and sink.

An undercounter sink should be installed with clips holding the sink to the countertop above it. The sink is then removed by undoing the clips and lowering the sink into the cabinet below. With a ceramic tile counter, salvaging the quarter-round tile pieces around the top of the sink will be touch and go.

Sink Materials

There are five different materials used in kitchen sinks.

- **Cast iron**—This is the most common. Its enamel coating is hard and resistant to scratching and staining. As the surface becomes abraded with use, it becomes harder to keep clean and gives the sink a finite life. Items dropped into the sink can chip away the enamel, leaving an unattractive black spot from the cast iron underneath. The heavy weight of a cast-iron sink can be a problem in some mountings.

Available styles are made for drop-in, tile-in, or undercounter mounting.

- **Porcelain**—The major objection to these sinks is their relative fragility. They can shatter when something heavy is dropped in them.
- **Stainless Steel**—This common sink material obviously doesn't chip or stain. Its brightly polished surface does become scratched and less attractive with age. Where water is hard and a drop of water dries, the inevitable spotting shows up on any dark surface but especially on steel. Aesthetics are another objection to stainless sinks—some people just don't like them.

They are used for both drop-in and undercounter mounting.

- **Solid Surface**—The advantages of solid surface materials apply to sinks as well as countertops. These sinks are mounted under the countertop and, when properly made, the countertop and sink look like a single piece. They can be aesthetically pleasing and functionally top notch. While resistant to staining, the sinks can be scoured if necessary. Similarly, scratches can be burnished out because the material is the same throughout the whole sink body, not just on the surface. The major objection is the cost.

Solid surface sinks are designed for mounting under a solid surface counter.

- **Quartz Resin Composite**—These sinks are made by a number of manufacturers and are available in designs which range from plain old sinks to the exotic. The material is a mixture of quartz, a hard mineral, and a resin. Granite is sometimes used instead of quartz. Like the solid surface materials, composite sinks are the same throughout. The material is hard with the appearance of an enameled sink but, because there is no surface coating, it does not chip off. It is warranted against staining by most manufacturers. Scratches and tough stains can be rubbed out. Sinks made of this material are much lighter in weight than cast iron and some brands and styles are price competitive.

Composite sinks have been used in Europe and have been available in the United States for over fifteen years. Like everything else in home building they have been slow to gain popular acceptance. They are available for both drop-in and undercounter mounting.

Accessory Mounting

You may want to mount a soap dispenser, a filtered water tap, or an instant hot water spigot in your sink. For cast-iron sinks you are better off if you think ahead and get a sink with the holes cast in it. Holes can be drilled in cast-iron sinks but it's a tricky business in the presence of the hard enamel surface.

Stainless, solid surface, and composite sinks can be cleanly drilled for accessories when you want to add them.

Countertops

What you have in your kitchen is very much a matter of personal choice although there are functional differences among the different materials. Here are some things you should consider in making your choice of countertops:

• Appearance.

• Cost.

• Resistance to day-to-day wear.

• Resistance to cutting, scratching, and staining.

• Resistance to heat.

• Hardness. A hard surface resists cutting, scratching, and wear but will crack or chip if mistreated. Hard surfaces are more likely to cause breakage to glass or china objects dropped on them.

• Sink mounting. When sinks are mounted flat with or below the counter surface, debris can be wiped directly into the sink.

• Day-to-day maintenance.

• Repairability.

• Water lip. A lip along the front edge of the countertop that keeps minor spills from running on the floor.

Different countertop materials include plastic laminates, tile, stone, solid surface, and solid-surfacing veneer.

Helpful Hint. If you want an impressive kitchen counter at a price less than solid surface materials, consider stone. As explained below, it is handsome, easy to maintain and lends itself to undercounter sinks.

Laminates

The least expensive countertops are the laminates. These are made of a thin, hard, tough plastic sheet that is glued (laminated) to a piece of backing material, usually particleboard, for strength.

There are two major advantages to laminates: they are the least expensive of the countertop options and they are highly resistant to staining.

Laminates are not as hard as the other materials. This is a disadvantage in that they are easier to scratch or cut, but an advantage because dishes or glasses dropped on a laminate counter are not as likely to break.

Neither tile-in nor undercounter sinks can be used with laminates.

Laminates are relatively easy to clean but scratches will hold dirt. They are easily scorched with hot frying pans, cigarettes or even candles. They cannot be repaired except by replacing a section of the counter.

Seams in laminates are usually obvious.

Improperly manufactured laminate counters may separate from their underlying material making a lump in the surface. This should not occur but does occasionally.

Laminates come in many colors with and without patterns. There are two basically different ways to make backsplashes and counter fronts:

1. The countertop, the backsplash, and the counter front are formed from a single piece of material. There is a raised lip at the front edge of the countertop.

2. Only flat pieces are used, with the backsplash and front edge being separate pieces.

Aesthetically the one-piece counter is plain and utilitarian while using three-pieces allows for a much greater variety in design and appearance.

One-Piece Countertops—From the user's point of view, having the backsplash and the lip molded as a single piece has two distinct advantages: 1) There is no seam where the top meets the backsplash nor where it meets the front piece and 2) the front lip helps keep spills from running on the floor.

There is also a disadvantage. At inside corners there will be a 45-degree seam that is 34 inches long, going from the corner at the front of the counter to the back corner of the backsplash. Laminate seams are always utilitarian and this long seam is not a thing of beauty. Further, it takes special machinery to do a precise job of making this kind of countertop and not all fabricators have it. It's a tricky business, at best.

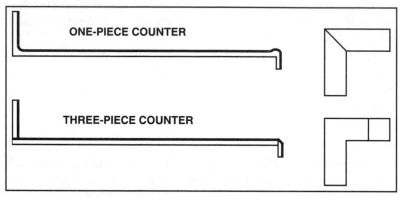

Laminate countertops can be formed as a single piece with unattractive corner seams or made in three pieces with less objectional seams. The 3-piece laminate counter is more popular.

Three-Piece Countertops—The inside corner problem can be avoided when the countertop is made with three flat pieces of laminate: the countertop proper, the backsplash, and the front. By cutting the corner section out of a single sheet, seams, if any are needed, occur in the top piece only and go in a straight line from front to back of the counter. Virtually anyone can make counters in this way and a lot of fabricators do. Sometimes tile is used for the backsplash to add a little variety. Other times wood is used for the same purpose. Counter fronts may be laminate or wood. To break the monotony, front edges may be beveled and faced with either the laminate or with wood. The variations seem to be limited only by the imaginations of the counter makers.

Note, however, that the three-piece laminate counter does not have the option of a lip on the front. Again there is a trade-off between form and function.

Ceramic Tile

Ceramic tiles come in many sizes, shapes, colors, and grades. In some areas tile has developed a reputation for being difficult to maintain and is seldom used even in upscale homes. This undeserved reputation comes from ignorance. I should know, I was one of the uninformed.

Our last three homes had ceramic tile kitchen countertops. We had no problems with either stains or pot marks with the first two. In the third we couldn't understand why the tile stained and why metal pots and pans made marks on the tile. It was while doing research for **Build It Right!** that I learned that matte finish tile is susceptible to staining and to pot marks. When I took another look at our kitchen tile, the light dawned— our kitchen counter tiles have a matte, not glossy, finish. We like the looks of the matte finish—but we're paying the price for it.

This is another place where you'll have to choose between aesthetics or function. My recommendation: if glossy surfaces turn you off, get a laminate or solid surface counter. Either that or be prepared to put up with stains on the tile from such things as fruit juices or wine (they will come out with bleach) and expect to live with putting a cloth on the countertop before you set anything on it that can make marks.

Glossy surfaces show scratches easier than matte finishes. The impact of this can be reduced by sticking to light colors in which the scratches and abrasions are not so obvious.

Tiles are glued to the underlying material with spaces left between them. These spaces are then filled with grout. Unlike the tile surface, grout is a relatively porous material and absorbs stains easily. This problem is handled in three steps:

• Have the tiles set as close together as possible, making a grout line which is in the order of one-sixteenth inch wide. Even if stained it's not obvious.

• Choose a dark grout—it won't show the stains as much.

• Use sealer on the grout. Depending on the type of sealer (and how much you pay for it) sealing may last up to 10 years. (Don't use a surface sealer, it may make the grout look shiny like the tile, but it doesn't penetrate and will need frequent reapplication.)

Tiles are resistant to knife cuts (the usual result will be to dull the knife) but it is possible to scratch the hard surface. This is repaired by replacing the tile. Be sure to have a small supply of every tile piece you have in your house so that it can be replaced if needed.

Also keep a record of the manufacturer and color of the grout. Thoughtful tile installers will leave a piece or two of each tile shape.

Tile surfaces are harder than laminates or solid surface materials. They have virtually no give so that a glass or a dish dropped on them will break easier than with other counters.

As discussed earlier, tile counters lend themselves to mounting the kitchen sink on top of, level with, or under the countertop.

The front edge tile (called a "V cap" because of its shape) has a raised lip to stop water from running off.

Aesthetics—Sometimes the backsplash will use a special decorative tile with a pattern to give the counter a more appealing look. Patterns can also be made with ceramic tile by installing them at 45 degrees to give a different-looking surface. This ability to have patterns in the overall tiled surface as well as in individual tiles is one of the attractions of this material.

Did you ever notice tile installations where the tile pieces along the edges are a different length than the rest so that there are two sets of grout lines, one for the edges and a different one for the rest of the counter? This doesn't have to be. It's just thoughtless, sloppy work.

Counters usually use the same tile for the backsplash. Rounded pieces give a finished look to the top row. They come in many different sizes, with 4¼- and 6-inch square tiles being common. If you have a backsplash with 4¼-inch tiles, it is not unusual for the installer to finish it off with 6-inch rounded trim pieces. This results in the two sets of grout lines. This has been done so often that some people don't object to the appearance—but to many of us it is very amateurish. If trim pieces are available the same size as the tile,

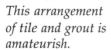

This arrangement of tile and grout is amateurish.

tile, they should be used, otherwise cut them so that the grout lines line up in a neat and orderly fashion.

In one unfinished house I visited, an installer was putting tile in a tub surround and the edge pieces weren't lining up with the tile. In this case the tile was 6×8 inches and the edge pieces were 6 inches long. However, the way the tile was laid, the 6-inch edge pieces were finishing the 8 inch side of the tile! When I asked if he didn't think it would make a nicer looking job with the grout lines lined up, he said that, no, he liked the random appearance of the two sets of grout lines!

To each his own, but if you want your tile job to look as if the tile and the trim pieces came from the same place, it is suggested that you make this desire known to the builder and the tile installer. Be sure, too, that tiles and trim pieces of the same size are available in the tile pattern you select. It's a good investment to have it done right.

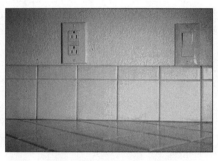

Coordinated backsplash tiles.

Workmanship—Good and bad workmanship in tile installations is usually obvious; tiles don't lie flat, grout lines aren't straight, etc. Proper edge-cutting is another indication of the quality of the work. Inevitably there will be places where the installer will have to cut a piece of tile or trim to make it fit the available space. These cuts, whether made with a saw or made by scribing and snapping (breaking) the tile will leave edges that are sharp and, sometimes, rough. A tile-cutting saw makes cleaner cuts than does a snap cutter.

These edges can be smoothed and rounded by using fine sandpaper or a grind stone. The color of a piece of ceramic tile is in the surface glaze so that, when the glaze and the body are distinctly different colors, rounding the edges too much lets the underlying color show through. This should never be a problem if a saw is used to cut the tile. Rounding is then only necessary to smooth the sharp edge—and not to smooth away chips.

Another sign of poor workmanship—or rather poor job management by the builder—is to have an electrical box partly overlapping the backsplash tile. It should be totally inside the tile area or totally outside of it. The arrangement shown here is not necessary and there's no reason you should have it in your new home.

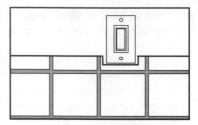

This results from poor supervision of subcontractors.

Maintenance—Sealers to protect grout from staining have to be renewed periodically. In selecting a grout color, remember that light colored grouts are more likely to show stains and discoloration than darker ones. Your supplier should tell you this, but it's sometimes overlooked. If it can be worked into your decor, choose a darker grout color because it's easier to maintain.

This is an area where improvements are always being made. Check with a competent tile supplier to see if there may be a grout available that is inherently resistent to staining.

Tile won't burn but care should be exercised about putting something very hot on it because it can crack from thermal stress.

Stone

Stone usually comes in the form of large tiles and some of the pros and cons of ceramic tile also apply to stone countertops. But since the material is the same color all the way through, it is possible to work out surface scratches or even shallow chips. The cautions about cutting apply here also.

Granite makes an attractive, durable countertop.

Space between pieces of stone are kept very thin and grout lines are not a source of concern for staining. As noted earlier, a stone tile counter is more expensive than tile but the improvement in appearance can be dramatic.

Granites come in a surprisingly large range of colors and patterns. Since these are not just on the surface it is possible to shape the edge of a piece of stone. This is particularly useful around edges of an undercounter sink and no special edge pieces are needed as they are with ceramic tile.

Sometimes a fairly large area will be covered with a single piece of stone (one lady wanted a large smooth surface to roll out crusts for her baking). It is dramatic since the effect cannot be achieved in any other way. And it is expensive, both for the material and for the installation.

Solid Surface Materials

There are a number of different countertop materials in the solid-surface category, Corian being the best known. These surfaces are two to three times more expensive than ceramic tile. Solid-surface countertops have a smooth surface that is almost impervious. It is stain resistant and it doesn't have grout to keep clean. Properly made seams are virtually invisible. Sinks of the same material are available that can be mounted under the countertop so that they seem all one piece.

Solid-surface materials have the same consistency throughout. They are a cast resin that may be polyester or acrylic or a combination of both, giving somewhat different properties to the resulting materials. Generally all-acrylic counters, like Corian, are more durable and a little more expensive.

Maintenance—Durability is excellent. Burns can be polished out as can cuts or scratches.

Some solid-surface counters come in a matte surface while others are glossy. Glossy surfaces will show minor scratches more than matte finishes, but these can be buffed out. It does mean more maintenance, however.

Uses—Because the joints can be buffed smooth, two pieces can be directly abutted to make invisible seams. It is this property that makes it possible to cut out damaged material and replace it without the repair being obvious. It can also be seamed with other materials such as wood, brass, etc. It's the material of choice for kitchen countertops—or any other application that requires a tough smooth surface that is visually appealing.

The primary drawback to solid-surface countertops is the price, but if you can afford them, they're both elegant and utilitarian. They are used in kitchens because their greater resistance to cuts and scratches and their repairability make them more attractive than other materials. Solid surface materials are shipped by the manufacturer, usually in flat sheets, to fabricators. The extensive seaming, cutting, and grinding operations needed to make them into counters are done locally.

Solid Surfacing Veneer (SSV)

A recent innovation is solid surfacing veneer made by Wilsonart. Older solid surface materials are one-half inch thick, solid all the way through. The all-acrylic solid surfacing veneer is one-eighth inch thick and is laminated on top of a different, less expensive sheet material, like particleboard, for strength and rigidity. The combination has the advantages of the solid surface material for appearance and maintainability but at a significantly lower cost. SSV counters are, however, still somewhat more expensive than tile.

Solid surface sinks of the same material as the SSV are available but their cost detracts from their attractiveness. Because of the weight of cast iron undercounter sinks, they are difficult to mount under an SSV countertop. However, there is no obvious reason that lighter weight and less expensive sinks, stainless steel or quartz resin composite, could not be mounted under an SSV counter to make a handsome, functional, and relatively inexpensive combination.

You will probably find a reluctance of fabricators to do this, however. Until the SSV has been in use for a few years and there is more experience with it, shops are not going to put their necks out. Their reasons:

- First, all solid-state manufacturers are very wary of using their materials in a fashion that could jeopardize the reputation of their products. Therefore, the only "approved" sink to mate to an SSV counter must be manufactured by Wilsonart.

- Second, there have been past failures when two dissimilar materials were cemented together and SSV has not yet withstood the test of time.

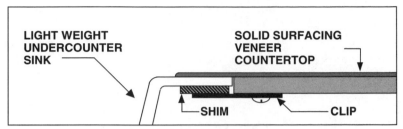

Undercounter sink clipped under a solid surfacing veneer counter.

• Third, until more experience has been gained, there are serious doubts about how much less the installed costs really are compared to non-veneered solid-surface materials.

The availability of veneers for laminating brings the advantages of solid-surface material within just about anyone's reach. As this material becomes more widely used, we should expect to see a variety of both aesthetic and functional options.

Breakfast Bars

Many kitchen designs include an island or a peninsula that separates the kitchen from the family room or the breakfast nook. This island or peninsula is often designed to double as an eating bar. When you see such a plan be sure that the countertop extends out far enough to let someone sit there (12 inches is a minimum, more is better).

If the cooktop is also on the same counter as the breakfast bar be especially careful that there is a way to prevent a person using the bar from getting burned by something on the cooktop. (*See* the warnings in the section on islands in Chapter 19.)

Chapter 14

Dishwashers

The most important thing about dishwashers is how well they do their job and you can find out about that from your local appliance dealer and from *Consumer's Reports*. Our concern here is where the dishwasher should be located to minimize the work you have to do to use it.

The issue is not trivial—sometimes an otherwise desirable house plan has no suitable place for the dishwasher whose location is one of the key elements in a good kitchen design. How the dishwasher location interacts with the overall kitchen design is discussed in Chapter 19.

The requirements for the dishwasher location are few and simple—but, oh, so important for that good design that minimizes the effort needed to load and unload the dishwasher. Numbers in parentheses are from the floor plan study discussed in the Introduction; they give the fraction of new houses where a particular problem was found.

1. The dishwasher should not be separated from the sink. (19%)
2. The dishwasher should never be in a counter at 90 degrees to the sink. (20%)
3. When the dishwasher or the sink is at a 45-degree angle, the dishwasher and the sink should be between 6 and 12 inches apart.
4. When the dishwasher or the sink is at a 45-degree angle, there should be two cabinet doors under the sink or, if there is only one, it should be hinged on the same side as the dishwasher.
5. There should be plenty of room in front of an open dishwasher door to walk. You shouldn't have to go around an island to unload the dishes.

6. Most of the cabinets for the dishes, glasses, silverware and cook-
   ware from the dishwasher should be located within one step of
   the dishwasher door when it is open. (39%)

Dishwashers and Sinks

Two concerns about the separation between sink and dishwasher
are the potential of dripping water on the floor when loading the
dishwasher and the number of steps you need to take in the process.

With today's dishwashers, according to *Consumer's Reports* (Au-
gust 1995), dishes do not need to be rinsed as in the past; just scraping will do. But for many of us, "scraping" means rinsing them off
into the disposal without doing a thorough job. And the dish is wet
either way.

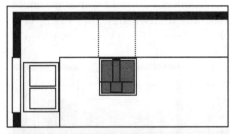

*Dishwashers and sinks
should not be separated.
The layout in the diagram
is bad, the one in the photo
is the worst.*

Besides this, when the dishwasher is separated from the sink by more than about a foot, you have to take at least one extra step to put dishes into the dishwasher. This, coupled with dishes that may be wet even when scraped, argues that the dishwasher should be right next to the sink. All in all, it's just more convenient to have it there.

Some how-to design books state emphatically that a right-handed person will find a dishwasher easier to load if it is to the left of the sink. I personally have had both kinds of kitchen layouts and cannot see what possible difference it makes. Check this out for yourself and, everything else being equal, put the dishwasher to the left of the sink but don't make it an imperative in your layout.

Dishwashers in Corners

When the dishwasher is in a corner, the open dishwasher door blocks access to base cabinets and makes access to upper cabinets either difficult or impossible. With the dishwasher door open, you should have easy access to *all* cabinet doors. When you do not, you'll

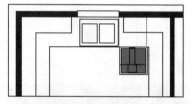

end up unloading the dishwasher on to the counter, closing the dishwasher door, and then putting the dishes away. This extra handling is not only extra work, but is a good way to chip dishes and glasses.

The problem occurs most often when the sink is at the end of the a narrow U-shaped kitchen. In any case, be sure there's a convenient amount of space between the dishwasher and the closest cupboard. You should be able to comfortably stand there while moving dishes from the dishwasher to the cupboards. Two feet is a good value but a little less can be used if necesary. And be sure that you can get into the base cabinets that are beside the dishwasher.

A dishwasher in a corner is a sure sign of an inept design.

Seriously consider rejecting an arrangement where there is no room to stand beside both sides of the dishwasher or if it isn't possible to open cabinet doors when the dishwasher door is open.

45-Degree Sinks

Anyone working in the kitchen appreciates lots of windows and airiness around the sink and the adjacent work areas. When the layout of the house permits, having the sink across a corner with windows on two sides is attractive and makes life in the kitchen more pleasant. There are a couple of precautions, however, that should be observed:

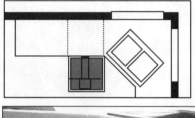

- When the sink is set at 45 degrees across a corner, be careful not to have the dishwasher too close to the sink. If there is no space at all, an open dishwasher door will create a narrow corner in front of the sink that makes it difficult to move when you're loading the dishwasher. A spacing of 6 to 12 inches will solve the problem. Don't go more than 12 inches or it'll mean extra steps and dripping when loading the dishwasher.

- When a single door is used on the cabinet under the sink, the door should be

Separate dishwasher and 45-degree sink by at least 6 inches so you have a place to stand.

hinged on the *same* side as the dishwasher, otherwise you will not be able to get to the garbage container under the sink when loading the dishwasher. Double doors, of course, do not have this problem.

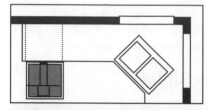

More than 12 inches between a 45-degree sink and the dishwasher will cause extra work and result in wet dishes dripping on the floor.

Be sure to check the hinge on a 45-degree cabinet door.

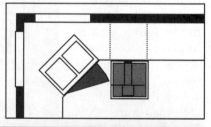

Dishwashers and Cabinets

One of the most annoying things a designer can do is to separate the cabinets from the dishwasher so that steps are required as each item is put away. Sometimes there is a single upper cabinet above the dishwasher and in other designs there is none at all. Two different arrangments account for most kitchens where this separation occurs:

- In one, the sink and dishwasher are placed on a peninsula where a person using the sink can look across the family room or breakfast nook to a view or onto a golf course. There are better ways to take advantage of the view. If you don't want to take many extra steps unloading your dishwasher, avoid sinks and dishwashers on peninsulas unless there are cabinets over the peninsula.

- In the second arrangement there is a big window with the sink centered on the window and the dishwasher is placed on the side

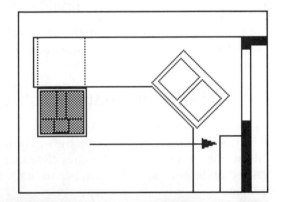

A dishwasher on a peninsula is usually a problem. Here it is separated from both the sink and the cupboards.

A dishwasher should be next to cupboards, not across the room.

of the sink away from the cabinets. Sometimes this is needed to keep the dishwasher out of a corner and in others it's just a plain bad design.

Space in Front of Dishwashers

Open dishwasher doors stick out 24 inches into the aisle. At least 18 inches should be left to get by the open door, more if possible. This means a minimum of 42 inches clearance in front of the dishwasher, preferably 48 inches. An island is the usual reason for the problem.

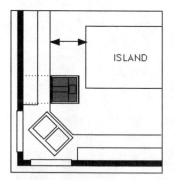

ISLAND

Allow at least 42 inches in front of a dishwasher, 48 inches is better.

Chapter 15

Refrigerators

The location of the refrigerator in the kitchen is important in a user-friendly home:

- You should not have to take extra steps because of its placement.
- You should be able to open all the doors far enough that shelves and internal bins and trays can be removed.
- You should not have to walk around an open door.
- You should be able to stand in front of the refrigerator, take things out of it and have a place to set them without moving your feet.

In many kitchens the refrigerator location does not meet these criteria, at least not for all refrigerator configurations. Refrigerators are not a minor cost item. When the day comes for you to sell, you will find that people will object to having to buy a new refrigerator if they buy your house. It is important, therefore, to have a refrigerator location that is suitable for all refrigerators. And don't forget that you may want to update your own refrigerator someday and it should be feasible to change from a top-mount or a bottom-mount to a side-by-side model, or from a deep model to a thin one.

These considerations are the subject of this chapter. We start with a review of refrigerator configurations and sizes.

Refrigerator Configurations

Refrigerators may be free-standing or built-in. All have two compartments; one is for fresh foods and one is a freezer. The location of the freezer gives rise to the descriptions of "top-mount," "bottom-mount" and "side-by-side" units. Built-ins are usually side-by-side units but they may have separate freezers that are similar to the pull-out freezer of a bottom-mount refrigerator.

When doors don't open fully, the shelves on the doors are in the way when you need to pull out the bins or shelves inside the refrigerator. This, in turn, limits access to what's inside and also means that bins and shelves cannot be removed for cleaning.

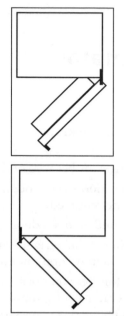

Top-Mount and Bottom-Mount Models

In top-mount models the freezer is behind a hinged door the full width of the refrigerator. Both the freezer and fresh food doors swing from the same side. Access to foods at the back of the freezer compartment is not convenient.

In bottom-mount models the freezer is a pull-out unit the full width of the refrigerator. Again access to frozen foods is not convenient.

Our concerns here are the problems that can occur when the opening of doors is restricted and, in this respect,

With top-mount refrigerators, hinges may be on either side.

both top-mount and bottom-mount models have the same potential. To simplify the discussions, they are grouped together and called top-mount in the rest of the chapter.

Most top-mount refrigerator doors can swing from either side by simply moving the hinges. For models with fixed hinges you have to decide which side the hinges should be on before you buy the unit.

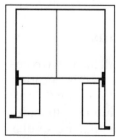

Side-by-Side Models

Generally, side-by-side are more expensive than top-mount units having the same size ratings. The availability of cold water and ice through the door is an attractive feature and. when frozen food is kept in bins that can be pulled out for access, side-by-side units are also more convenient to use—if they're located

Side by side model. properly.

In side-by-side units the two sections of the refrigerator mount vertically with separate doors with the freezer on the left when facing the refrigerator. On most refrigerators the hinges are arranged so that, when they are opened straight out, the fronts of the doors align with the sides of the unit.

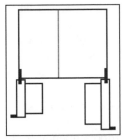

Large side by side. Thicker doors make the difference.

In a few of the larger side-by-side refrigerators, hinges are mounted differently and a significant part of the rather thick doors swings out past the side of the box. With at least one major manufacturer, the difference between sizes of 27 cubic feet and 25 cubic feet is not the box itself but in the doors. By adding a little over an inch to the door thickness the inside volume is increased enough that they are rated at 27 instead of 25 cubic feet . Note that this adds nothing to the internal shelf space—in fact the shelves are interchangeable in the two units. It's all in the door thickness. To keep this thicker door from blocking even more of the box when it is open, the doors are hinged differently and do not line up with the edge of the box when they're straight out.

A style that is relatively new is the "thin" unit that sticks out past the front of the kitchen cabinets only far enough for the doors to open. These can be mounted to have a "built-in" appearance. Don't be in a hurry to get one, however. For the same number of cubic feet of storage they are significantly more expensive and wider than deeper units. They are also likely to have shelves, not bins, in the freezer, making access to frozen foods even more difficult.

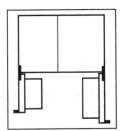

With a thin side by side, the body isn't as deep.

Refrigerator Dimensions

When looking at refrigerators in appliance stores, one gets the definite impression that manufacturers try to outdo each other to make a unit that is a different size than any that has existed before. The dimensions listed in the table are the result of a survey of a large number of units. Undoubtedly units exist which do not fall in the ranges listed so use the table as a guide to give you some idea of what to expect.

Approximate Refrigerator Dimensions

Side-by-side: **29.8 cubic feet**
  Height   69¾   inches (includes hinges)
  Body width   35¾   inches
  Overall width   38   inches (doors open at 90°)
  Body depth   28½   inches
  Overall depth   34½   inches (including handles)
  Door thickness 3 ¾   inches (excluding shelves)

Side-by-side: **25 to 27 cubic feet**
  Height   68¾   to 70½   inches (includes hinges)
  Body width   35¼   to 36   inches
  Overall width   35¼   to 38¼   inches (doors open at 90°)
  Body depth   27½   to 29¾   inches
  Overall depth   31¾   to 36¾   inches (including handles)
  Door thickness 2¼   to 4⅝   inches (excluding shelves)

Side-by-side: **22 to 25 cubic feet**
  Height   68¾   to 70½   inches (includes hinges)
  Body width   35¼   to 36   inches
  Overall width   32¼   to 35¼   inches (doors open at 90°)
  Body depth   26⅝   to 28½   inches
  Overall depth   32½   to 35¾   inches (including handles)
  Door thickness 2¼   to 3½   inches (excluding shelves)

Thin side-by-side: **20 to 24 cubic feet**
  Height   69¾   inches (includes hinges)
  Body width   35¾   inches
  Overall width   37⅞   to 38¼   inches (doors open at 90°)
  Body depth   23¾   to 23⅞   inches
  Overall depth   27⅝   to 28⅞   inches (including handles)
  Door thickness 2¼   to 2½   inches (excluding shelves)

Top-mount: **19 to 25 cubic feet**
  Height   64¾   to 68   inches (includes hinges)
  Body width   31   to 34½
  Width   31   to 32¾   inches (door open at 90°)
  Body depth   26⅝   to 28½   inches
  Overall depth   31½   to 32⅞   inches (including handles)
  Door thickness 2¼   to 3½   inches (excluding shelves)

Refrigerator Placement

With shelves mounted on the doors of all refrigerators, it is necessary to open the doors well past straight out to be able to pull the interior bins completely out. Since these will need to be removed for cleaning from time to time, it is important that doors be capable of opening as far as they're designed to. This is approximately 150 degrees—not all the way around to the side but well past straight out front.

One manufacturer's catalog says that you can get the fresh food pans out even with the door at 90 degrees—if you remove the door bins first. *Try it for yourself. On the units I looked at this wasn't so; you have to open the doors way past 90° degrees to get the interior bins out—regardless.* For side-by-side freezer doors, this ability to open fully is also needed to access the ice-cube storage area.

For at least one model of the thin units, full access to the freezer is possible with the door at 90 degrees. This is done by limiting the amount of door storage and by using wire shelves rather than pull out bins. These shelves do not pull out, thus severely limiting access to whatever may be stored on the back of the shelves. When such a unit is mounted as a built-in, there is no longer a problem with someone wanting to put a different model there in the future. If the unit is not built-in, it may be a serious drawback to anyone else wanting to buy the house in the future.

Put simply,

> **There should never be anything next to a free-standing refrigerator that restricts how far any door can open.**

Refrigerators Next to Walls

With the exception of a stub wall beside the refrigerator used for cosmetic reasons, putting the refrigerator next to a wall will cause a problem.

For side-by-side models, a wall will stop a door from opening more than straight out so that you can't get internal bins out.

With top-mount models the doors could be hinged on the side away from the wall to avoid this problem except that now you would have to walk around such a door every time you use the refrigerator.

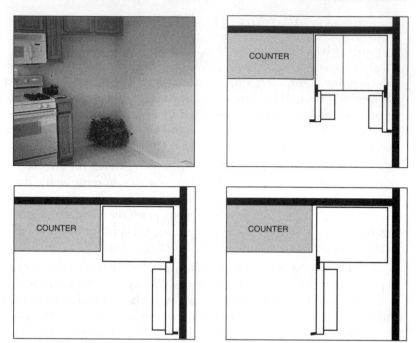

Don't put a refrigerator beside a wall. A door won't open fully or you'll have to walk around one.

The safest thing is to go with a kitchen plan in which the refrigerator does not have a wall on either side.

In unfurnished models, sales agents will sometimes put pictures or potted plants into the holes where the missing refrigerator would go. When you are looking at models be careful to not be distracted by these. It's still a bad location.

Refrigerators Next to Doorways

A fairly common arrangement is to put the refrigerator beside the wall that has the door into the dining room. As shown in the graphics, this may allow the doors to swing open farther but it doesn't solve the problem.

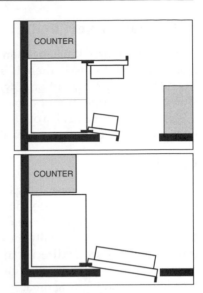

A door won't open fully when the refrigerator is tucked into a corner beside a doorway.

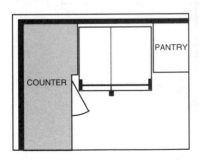

A different bad corner for a refrigerator.

Refrigerators Next to Right-Angle Counters

There are worse kitchens than those which put the refrigerator against a wall! In these, there is a right-angle counter right beside the refrigerator. Now not only will the adjacent door of the user's side-by-side refrigerator not open but the cabinet drawer and door under the counter will also open only a little way, making the whole back end of the under-counter cabinet virtually useless.

Landing Areas

When you are putting things into or taking them out of the refrigerator, you need a place to set them that is within easy reach when you are standing in front of the open refrigerator. The "landing area" that works equally well for top-mount and side-by-side units is counter space directly in front of the unit. For top-mount models counter space next to the latch side of the refrigerator also provides a good landing area.

To reach a landing area you should not have to reach around an open door and it's more convenient if you don't have to reach across the closed door of a side-by-side unit. From this we come to the following conclusions:

1. It is better to have a landing area right in front of the refrigerator. This works well for any type of unit.

2. If the kitchen layout doesn't permit a landing area in front, a top-mount unit will be more convenient to work with. (Of course you have to trade this off against the overall inconvenience of the top-mount unit.)

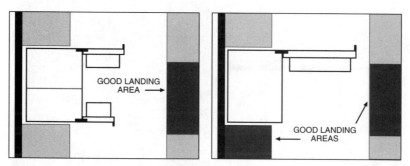

Good landing areas make it easy to use the refrigerator.

Screening Walls

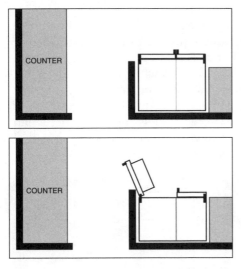

A screening wall used to hide the side of a regular refrigerator may not be suitable for a thin one.

When a refrigerator is turned so that its side is beside the doorway, a screening wall can be added beside it to give a more finished appearance. It is strictly cosmetic but, if not done properly, it *can* interfere with function. If the wall is too long it will stop a door from opening. And particular care needs to be exercised if one of the thin models *may* be used because the wall may have been designed for a regular model and will be too long for the thin unit. 25 inches is a good length for the screening wall beside a refrigerator.

Door Clearance

When an island or other counter is too close to the front of a refrigerator, it forces the use of a side-by-side model with its smaller doors. This is generally not a good design because not everyone will be happy about this constraint when resale time comes. A good distance is at least 64 inches from the wall behind the refrigerator to the island or counter in front of it. Of course, if you are planning on a built-in side-by-side refrigerator, this distance can be less.

Be sure to allow adequate clearance in front of the refrigerator.

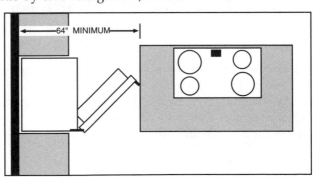

Refrigerator Depth

It is common practice in kitchen floor plans to show the refrigerator about the same depth as the adjacent counters. This can be misleading since refrigerators are frequently much deeper than the 24 or 25 inches of a counter. For a 27 cu. ft. unit, the overall depth (not including the door handles) is $31\frac{5}{8}$ inches, almost 8 inches in front of the counter. Sometimes this causes no problems but in others the refrigerator may be in the way or look out of place. Use your measuring tape when looking at models and your ruler when checking plans. (If plans don't have dimensions, remember that standard counters are 24 inches deep.)

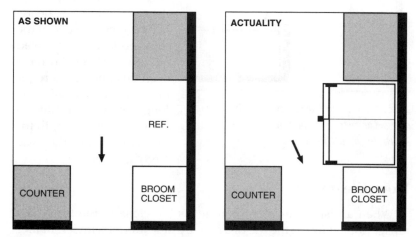

When refrigerators are missing in plans or models, it hides the problems they can cause.

Other Things to Watch For

Height

Take care to check out the height of the space for the refrigerator. As noted in the earlier tables, there is a wide range of heights, even for comparably sized units. If the cabinets are made for a $66\frac{3}{4}$ inch high refrigerator then there'll be real problems if someone comes along with a $70\frac{1}{2}$ inch unit.

Water

Most plumbers use a small recessed metal or plastic box behind the refrigerator in which to terminate the pipe for the refrigerator's ice maker water. This makes the connection more or less inside the wall but in one tract I visited the connector for the water is in front of the wall so that the refrigerator must be put several unnecessary and wasted inches farther out into the room. Be sure to insist on the recessed arrangement.

Some User-Friendly Refrigerator Placements

We have noted two major criteria in choosing a location for the refrigerator: (1) don't block a door from opening well past straight out and (2) have a suitable landing area, preferably in front of the unit. To these we add a third: keep the refrigerator close to the work area around the kitchen sink. There are several basic arrangements that meet these criteria. These are incorporated in the basic kitchen designs explained in Chapter 19.

Counter in Front of the Unit

In galley-type and narrow U-shaped kitchens the refrigerator is kept away from a wall that could cause door problems and the opposite counter is a good landing area.

Island

An island provides an excellent landing area for a refrigerator.

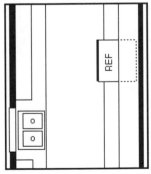 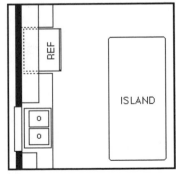

These arrangements are excellent for any refrigerator model.

Peninsula

When the refrigerator is opposite the end of a peninsula in a G-shaped kitchen you have a good landing place that can also be close to the sink work area. Putting the refrigerator here is also advantageous because it is then closer to the breakfast nook when serving and putting foods away.

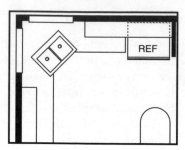

A peninsula can be useful as a landing area.

In-Line Layout

When there is no landing area opposite, a top-mount unit is needed to fully meet our criteria. If a side-by-side were used, you would have to reach across a closed door or go around an open one when moving food between the work area and the refrigerator.

When the refrigerator is in the same counter as its landing area, a top-mount unit is more convenient.

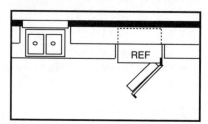

Chapter 16

Cabinets

Cabinets are an important part of how a kitchen looks and, to a lesser extent, how it functions. You will probably select your cabinets based on cost and appearance. But aesthetics and function are not totally independent and you will find it useful to review the functional aspects of cabinets before you make your final decisions.

Cabinet Styles

There are two distinctly different cabinet styles: western or face-frame and European or box. In western-style cabinets the front edges of the sides of the cabinet are covered by a ring of wood, the face frame. This ring is missing in European-style cabinets so that the cabinet looks like a box from the front.

Face-frame Cabinets

In a western-style cabinet the face frame is clearly visible around the top and side edges of closed doors. When two cupboards are adjacent, a common face frame piece is used. The width of the face frame varies from one cabinet to another, depending upon the design and manufacturer.

With western-style cabinets, traditional hinge units are visible from the front of the cabinet along the edges of the doors. These hinges are made of thin pieces of metal that mount to the door and to the side of the face frame. Doors open a full 180 degrees and, when open, the cabinet opening is obstructed only by the face frame and the thin metal of the hinge mounting.

The space behind the face frame is useful for storage when shelves don't pull-out. With pull-out shelves, however, this space is lost because shelves must be narrower than the inside of the face frame.

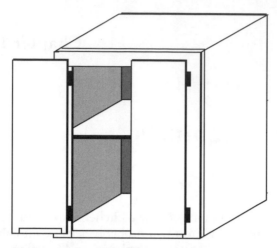

The western or face-frame cabinet with conventional hinges. Doors open all the way.

European Cabinets

Without the face frame on which to mount doors, the box cabinet differs in several ways from the face-frame design. Doors are mounted with hinges that mount to the insides of the cabinet behind the doors. The hinges are hidden and the doors don't swing clear of the cabinet opening causing a problem for pull-out shelves.

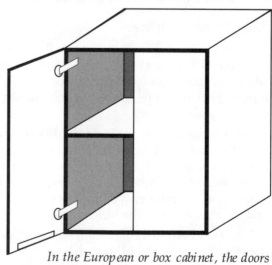

In the European or box cabinet, the doors don't swing much past straight out.

Cabinet Hinges

As with cabinet styles, the choice of hinge types is very much a matter of personal preference. Here is some background information about hinges to help you make your choice.

There are two general categories of hinges:

- Hidden hinges in which no part of the hinge can be seen with the doors closed. There are two distinctly different types of hidden hinges—the knife hinge found only on face-frame cabinets and the European hinge used on both cabinet styles.

- Visible hinges, where some part shows from the front. These are used only with face-frame cabinets.

Hidden Hinges for Box Cabinets

With box cabinets the doors must be adjusted carefully both for proper operation and for appearance. This is accomplished by using a style of hinge that mounts on the inside of the cabinet wall and which is easily adjustable sideways, in-out, and up-down. This is the European-style hinge.

With this hinge the door part of the hinge fits into a round shallow hole cut into the back of the door. When used on box cabinets, the hinge sticks back about two inches into the cabinet with feet that mount to the wall.

These hinges, while invisible when the doors are closed, are large and very visible when the doors are open.

The hinge pin itself is inside the hole cut into the door. As a result of this construction:

- Doors don't open much past straight out (95 to 115 degrees), and

- When open, the inside edge of an open door doesn't clear the cabinet opening thus reducing the effective width.

Having the door hinge this way is necessary so adjacent doors on a cabinet don't hit each other (remember that without a face frame, doors on adjacent cupboards are themselves adjacent).

Because the hinge mounts on the inside of the cabinet wall and open doors don't clear the opening, pull-out shelves in European-style cabinets have the same problem as face-frame cabinets—the useful shelf width is noticeably less than the inside width of the cabinet.

There are "double jointed" versions of this hinge that do let doors open all the way. These are larger and even less attractive than the normal hinge—but they are hidden (as long as the doors are closed) and they do let the doors swing open. They are used in places where there are special requirements, such as corner openings for lazy Susans.

Hidden Hinges with Face-Frame Cabinets

The knife hinge, used only on face-frame cabinets, has a notch cut in the edge of the door where the hinge is. This arrangement lets the door open fully while still being hidden. (In some cabinets with these hinges, the notch *is* visible from the front. Done properly, it should not be.) This hinge is invisible, it lets the door open fully, and it does not take up valuable space that may be needed for pull out shelves in deep cabinets. Because of its construction it looks flimsy (particularly when compared with the Mac-truck appearance of the European hinge) and cheap. Hence there is some resistance to using it in expensive cabinets—even though it has functional advantages.

The European-style hinge. You'll see several adaptations of this hinge used on face-frame cabinets:

- Hinges with mounting plates that fit unobtrusively on the inside edge of the face frame.

- Hinges with mounting feet that fit on the front of the face frame with the door part of the hinge closing over the mounting. This takes an unusually wide face frame.

- A version of the conventional European-style hinge where the mounting is at the inside edge of the face frame but the hinge sticks well back into the cabinet.

This last is a lousy design for two reasons:

1. The useful front opening of the cabinet is reduced in width by both the face frame *and* the thickness of the hinges.

2. The long side of the European-style hinge juts back behind the face frame leaving an unsightly and useless gap between it and the inside wall of the cabinet.

The result is that pull-out shelves must be even narrower and the unattractiveness of the European-style hinge is accentuated even further. The only good thing that you can say about the arrangement is that the doors are easily removable because the hinges snap together. Cabinetmakers can do a sloppier job with the doors and still use them because of the adjustment capability that is an integral part of the hinge. In other words, they are used for the convenience of the cabinetmaker, not for any benefit to the home user.

This awkward-looking European-style hinge on face frames is frequently used by custom cabinetmakers, but not by pre-manufactured cabinet makers.

There is a serious drawback to all of these European or pseudo-European hidden hinges, whether used in box or face-frame cabinets—they do not open much past straight out (95 to 115 degrees). Even a cursory examination shows that, when they are open as far as they'll go, there's so much leverage at the outside edge of a door that a small child could ruin a door or cabinet by pushing it. For cabinets above a kitchen counter this is more of a nuisance than a problem, but for base cabinets, pantries, or broom closet doors that go all the way to the floor, it is always a danger, particularly with a growing family.

When you're deciding about cabinets, remember that "new" and "in" aren't necessarily better. The older face frame cabinets may well be the best for you—they are certainly more traditional. When used with pull-out shelves, a cabinet with a minimum-width face frame is optimal for maximizing the available space.

Cabinet Exteriors

Cabinet exteriors offer many choices. The first decision is whether to use a wood (natural) finish or to paint them. If a wood finish is chosen, you need to choose the type of wood. Whether painted or stained, the finish color will be need to be selected. Generally, cabinets throughout the house are given the same finish but it is a matter of personal preference.

Cabinet exteriors range from plain to decorative. You may prefer finger grooves to open the doors or choose from a variety of hardware pulls.

Soffits

Should the cabinets go all the way to the ceiling? If not, should the space above them (the soffits) be boxed in?

When you are making these decision about boxing in the soffits, there are several things to consider:

1   Don't forget the vent for the exhaust fan. If the exhaust system is an updraft unit, it usually vents through a duct going up through the ceiling. This duct work will need to be covered up, either by boxing it in or hiding it with tall cabinets. If the system vents through a wall, there is obviously no duct that needs covering up. (But as a matter of good practice, ducts should be straight without bends to go through walls. *See* Chapter 17.)

2. Some folks like to have the space above the cabinets as a place to display knick-knacks. Others consider it a dust catcher and prefer that soffits be boxed in.

3. When the soffits are boxed in, differently colored paint or wall paper is sometimes used on the wallboard above the cabinets to add a decorative or festive touch to the kitchen.

4. A consideration with cabinets that go to the ceiling is that the top shelves are useful only for storing things you almost never need, because you have to use a chair or a ladder to get to them.

Tall Cabinet Doors

Here's a word of caution about very tall cabinets: wood warps and twists. Cabinet doors are no exception and, when they're warped, they are apt to rattle or bang when you close them—and gaping doors are not attractive either. Tall doors are more likely to do this than shorter ones. This problem has been seen in many tracts as well as in custom and spec homes in all price ranges.

Some cabinet manufacturers add stiffeners to the inside of tall cabinet doors. These will reduce warping but not twisting. Some cabinet makers put catches on these doors to overcome the effects of warping and twisting. These make opening the doors more difficult and add one more piece of hardware that can malfunction. Tall doors are, more often than not, unsatisfactory.

Cabinet Interiors

Particleboard

Particleboard is used in the cheapest cabinets. If the cabinets are not to be painted, two options are available: 1) a wood grain is printed on the surface or 2) a thin layer of wood is laminated to it. For painted cabinets this isn't necessary and the paint is applied directly to the particle board. Most people cover particle board shelves with a thin plastic lining or even vinyl linoleum to give a surface that can be easily cleaned and maintained. With better-quality cabinets this isn't necessary.

Melamine

Plastic laminate (melamine) is popular for the insides of cabinets and drawers because of its resistance to wear and its ease of cleaning. The basic plastic is the same material used in laminate (Formica) countertops and is glued directly onto the particleboard. Some cabinet makers will paint the shelf edges which doesn't do a particularly good job of covering the particleboard. Others use a vinyl strip that's attached to the shelf.

Usually this plastic is white but colors and even patterns are available. Besides the tough, easy-to-clean aspects of plastic laminate, it is popular with cabinet makers because it is cheaper and it doesn't have to be finished as wood does.

The edges of melamine-clad shelves and interiors should be smooth and clean. In the cheap cabinets seen in many tracts, the edges look like they had been cut with a hatchet.

Wood

Finished wood is found in better cabinets and drawer interiors. Good finishes on interior wood surfaces can be quite durable while poor finishes will stain, become dirty, and be difficult to keep clean. Well-made and finished wooden interiors will always have a richness in appearance that's not possible with plastic.

Backs

How the backs of cabinets are handled speaks loads about their quality. Cheap cabinets won't have a back. They'll have nailing strips for screwing to the wall and that's all. The underlying wallboard is exposed and the whole arrangement looks like it is—cheap. Better cabinets will be backed with the same material as the rest of the interior: wood or melamine. Look for the joints where the cabinet back and the sides meet. In well-made cabinets this joint will be smooth and tight. Sloppily-made cabinets will show definite, unattractive gaps.

Mountings

The nailing strip used to hang cabinets from a back wall is typically wood, plywood, or melamine-coated particle board. In good cabinets the material will have the same surface as the cabinet back and walls so that it matches the rest of the cabinet interior. Particularly for the melamine-coated particle board, this may result in a visible unfinished edge on the nailing strip that should be covered or painted to match the melamine but sometimes is not. The mounting screws used in better cabinets are not visible but are covered by special plastic buttons. (These are inexpensive and easy to use. Good cabinetmakers put them in place when the cabinets are installed. The who-cares installers don't bother. Actually it's something you can do yourself but shouldn't have to.)

Pull-Out Shelves

Center Mullions

Some cabinets have a piece of wood, called a center mullion, mounted vertically up the center of the front opening. The cabinet doors close against this mullion. This design is deadly to the use of pull-out shelves because it means you end up with two much smaller shelves. The mullions are in the way even without pull outs. Except in very wide cabinets, center mullions are used for cosmetic, not practical, reasons. If you're planning on pull-out shelves either initially or you plan to change over to them in the future, DO NOT choose a cabinet design with a center mullion.

In pantries and in other cabinets shelves may be adjustable up and down, they may be pull-out, or they may be fixed in place. When shelves are deep, typically 22 to 24 inches, articles in the back of them are difficult to reach. In these cases, pull-out shelves make a lot of sense. And shelves that are adjustable up and down, whether pull-out or not, also make a lot of sense. (It's not hard to make pull-out shelves adjustable, but you'll probably have to ask for them because cabinet makers don't usually do it.)

The amount of storage space is reduced when shelves are made pull-out because of the room taken by the slide hardware. However, the increased accessibility to the back of deep shelves will usually more than make up for the reduction in shelf width.

But there's more to it than just gettng pull-out shelves. In a subdivision that typifiies the builder who doesn't pay attention to what his subcontractors do, I saw lots of pull-out shelves! Below the counter, a 30-inch wide cabinet has a pair of double doors on the front. Instead of having the doors simply meet at

*Center mullions
don't go with
pull-out shelves.*

the center of the cabinet, there is a center mullion some two
or three inches wide. The doors close over this. And there are
two sets of pull-out shelves in the cabinet, one set behind each
door!

Even without being pull-out, the shelf's usefulness would be severely limited by the mullion. With the space lost for the two
sets of pull-out hardware plus the space taken by the center
mullion in front, close to half of the space that should have
been available for storage is lost.

Beware of pull-out shelves in which the front of the pull-out shelf
hardware is not covered. Regardless of how careful you are, sooner
or later someone will pull out a shelf without fully opening the
cabinet doors. When the front of the pull-out sliders is left uncovered, this will result in gouging the interior finish of a partly-opened
door. Some cabinetmakers put a piece of plastic in front of the hardware so that the plastic hits the door first, virtually eliminating the
problem. The usual arrangement has the front of the shelf extending out to cover the hardware. But some custom cabinetmakers
aren't up to speed on these things. (This problem was seen on both
coasts. It's not common but it does happen.)

The simplest pull-out design is just the shelf with slides mounted
on each side. These have no sides or backs. Then there are those
with fronts, sides, and backs several inches high, looking more like
a parts bin in a hardware store than a shelf. These are sometimes
called trays. Unless someone is expected to slam them around, there

seems little reason for other than minimal front, side and back pieces, if any at all. Such pieces take away from the space available for storage as well as add to the cost. But, again, personal preference should be the determining factor. If you have a choice, get what satisfies you.

While pull-out shelves can usually be added to an existing cabinet, this is not always the case. In some house designs the cabinets are irregularly shaped, making pull-out shelves impossible. Since these are usually deep cupboards, the home users are forever saddled with a large amount of virtually unusable cabinet space.

Lazy Susans

In corners of kitchen cabinets there will often be locations where fixed shelves leave areas that are almost inaccessible. In these cases a lazy-Susan type of arrangement is usually appropriate. These are more expensive than fixed shelves but end up providing more useful shelf space. These are available for below-the-counter cabinets in a number of different styles. Lazy Susans can also be used in the upper cabinets in kitchens although they're not as popular because the shelves aren't as deep.

An alternative to a lazy Susan in the corner is to have a door that is hinged in the middle with one half of the door along each side of the corner. The door then opens up nicely to give reasonable access to the shelves all the way back into the corner. There are four advantages to this arrangement:

• No space is lost from putting a round peg (the lazy Susan) into a square hole (the cabinet corner). All the space can be used albeit that farthest from the door is not as accessible as it would be with a lazy Susan. There is 20 percent less room on a lazy Susan than on a shelf.

• You don't have to spin the lazy Susan around looking for the pot or pan you want. It's all spread out in front of you.

• There is no problem with things falling off behind the lazy Susan. (When this happens you have to completely unload it to retrieve whatever fell off.)

• It's less expensive.

Don't forget, however, that to use this approach it is necessary that the opening be large when the doors are opened. A single door on one side of the corner won't do it. There are attractive arrangements where the door and the shelves inside are curved and no special hinges are required to open the door fully.

Be careful of door arrangements, whether for lazy Susans or just shelves in the corner, which involve two doors rather than one hinged one. While less expensive, these require that you open and close the doors one after the other. They are not convenient.

Drawer Boxes

Because cabinet drawers by themselves are similar to uncovered boxes, they are often referred to as drawer boxes.

The sides of drawer boxes may be connected to the fronts and backs of the drawers in two different ways. In one, the pieces are dovetailed, i.e., they are cut and grooved so that they intermesh at the corners. In the other, the pieces are simply butted and then stapled, nailed, or screwed. In all cases the pieces are glued. Dovetailing is "better" in that it has a more handcrafted look and is more expensive. Because of the glue, both approaches are more than sturdy enough for cabinet drawers. In a very expensive house, expect dovetailed drawer boxes; in starter homes expect stapled pieces. It's a matter of cost and taste. Utility doesn't enter into it. (Dove tailing is used only with wood, not with melamine or particleboard.)

Recycling Bins

A trend in cabinetry for today's environmentally conscious society is to have built-in recycling bins for used bottles and cans. They are a convenient amenity if you have room.

But don't do like one custom house I visited where the buyer carried this to the extreme and left out the disposal entirely! What isn't recyclable has its own bin and is carried to the garbage can. For her sake let's hope that when she gets ready to sell her house she can find a buyer who is as enthusiastic about not having a disposal as she was. Getting power under the sink and a switch on the wall isn't that easy after a house has been completed.

Finger Grooves or Knobs?

One of the more annoying things seen in kitchens with wall ovens—microwave or thermal—is that the doors on the cabinets just above the ovens come down too far. When knobs or handles are used, this is no problem. When the design uses finger grooves and the doors are too close to the oven, it is annoyingly difficult to get the doors open. The cabinet design should take into account that the top of the oven typically sticks out in front of the cabinet doors so that there must be enough space between door and oven to comfortably insert one's fingers—or use knobs.

To get the cabinets open, you need space for fingers between the oven and the cabinet doors.

Another place that finger grooves are very inconvenient is a bottom drawer where the grooves are only a few inches off of the floor. And cabinet doors above refrigerators are always hard to reach. Knobs or handles are a lot more useful.

The cost of putting knobs on drawers and cabinet doors is small. The inconvenience of not having them goes on a long time. They are always a safe bet if they go with the cabinet style.

Cabinet Bases

Watch out for custom cabinets in which the base, the part of the cabinet that sits on the floor, isn't mitered at the corners. The base is generally ¾-inch plywood with a hardwood surface. From the front of the cabinet the unmitered corners look fine but, if the cabinet is where it can be seen from the end, the view is of the edge of a piece of plywood, not the hardwood surface. Fortunately this sloppy piece of work isn't common but, if it does happen, don't accept it.

Custom versus Premanufactured Cabinets

Cabinets may be either premanufactured units from a large manufacturer or custom ones from a smaller, usually local, shop. Custom cabinets are generally perceived as more expensive and as of better quality than premanufactured units. This may or may not be the case. Actually you pretty much get what you pay for in both custom and premanufactured cabinets. Cheap is cheap regardless of who makes them.

One advantage of custom cabinets is that you can get them made exactly to size while premanufactured cabinets are available only in predetermined widths (usually 3-inch increments). When you are starting from scratch you can take this into account but when you're working with a kitchen whose dimensions are predetermined, it becomes more of a concern.

Local custom cabinet makers often don't finish the cabinets in their shops. This is done on site after the cabinets are installed. Generally speaking, the conditions for finishing the cabinets on the site are far from ideal. Temperature is not controlled nor is the amount of dust in the air. A principal reason that the small cabinet makers don't finish their cabinets in their shops is that they can't! There are stringent OSHA and environmental protection rules that have to be obeyed and it's expensive to equip a shop to do finishing. The rules don't apply when finishing is done in a house under construction.

The quality of finishing should be about the same regardless of where it is done—if it's done properly. Since there's less likelihood of getting a proper job on the site than in a shop, it's fair to assume that prefinished cabinets are the better choice if you have an option. (The same logic suggests that advertising which says that "site finished cabinets" are preferable may be misleading.)

You'll find that premanufactured cabinets are likely to be of more consistent quality. Some problems seen in custom cabinets, such as not leaving enough space between drawers for fingers, are unlikely to be found in premanufactured units. Further, with premanufactured cabinets, you can often have an exact preview of what you're getting. With custom manufacturers, you are at their mercy unless you can write out your specs in gruesome detail.

There is another consideration. A couple I know was held up on the completion of their new home for several months while they wrestled with subcontractors who were busy with more important jobs—meaning ones for people who had more clout. With premanufactured cabinets this wouldn't have happened.

It is important that the cabinets be installed properly. You have a right to expect this for both custom-made and premanufactured units.

A real problem with premanufactured cabinets is the question of how to get your kitchen designed properly. Retailers of these cabinets such as kitchen centers, home improvement stores, and larger hardware stores all use computer-aided design (CAD) systems that are impressive and efficient. You come in with your requirements and leave a half hour or so later with a completed kitchen design and a price for the cabinets. *But more times than not, that design will be faulty.*

Neither the CAD programs nor the training of the designers is directed toward the good kitchen design we are discussing here. Dishwasher, refrigerator, and other problems still occur. It is absolutely up to *you* to keep them out of your kitchen.

Chapter 17

Kitchen Exhaust Systems
(A.k.a. Ventilation Systems)

Exhaust systems, which are used to rid the house of steam, smoke, fumes, and odors, frequently exist because of a code requirement and/or because of a conception by the buying public that they should be there. Unfortunately, too often, in both kitchens and bathrooms, exhaust systems are ineffective in performing the function for which they were intended.

Cooktops are almost always equipped with a means for taking smoke, odors, and steam generated at the stove and exhausting it into outside air. Occasionally you'll see a house without a kitchen exhaust system or it'll be in the ceiling many feet away from the cooktop where it is useless.

Types of Exhaust Systems

Updraft—An updraft unit uses a fan in a hood or microwave oven that is mounted over the cooktop. It takes the air coming up from the burners and blows it through a duct to an outside vent, usually on the roof.

Downdraft—The fan in the downdraft units sucks the air sideways off the top of the stove and blows it down through a duct that goes outside.

In general, the downdraft units are not as effective in removing smoke, fumes, etc., as are the updraft units. It's natural for heated air to rise and not to be pulled down or sideways. To try to compensate for this, the fans for downdraft units are typically larger than for updrafts.

Another negative about downdraft units—pulling the air across the top of the cooktop will lower the temperature of anything cooking there. It is more noticeable with larger units than smaller ones.

Ductless—These systems do not exhaust the fumes but filter them and dump them back into the kitchen. They are found most frequently as updraft but may also be in a downdraft configuration.

Air Volume

Exhaust fans are rated by the amount of air they move and by the amount of noise they generate. In new homes you can often look under the hood and see the actual rating in CFM (cubic feet per minute). Typical updraft fans are rated at around 180 to 200 CFM. This appears to be reasonable for daily use. You can get bigger, more expensive, and noisier, fans. Downdraft ratings are from 300 to 500 CFM, although at least one manufacturer offers a unit that moves upwards of 1000 CFM. (For comparison, the rating of the fan in a forced air heating system in a 2000 square-foot house is typically about 1350 CFM.)

A word of caution—in today's tightly sealed homes very large fans have been known to pull soot down out of a chimney and spread it around in front of a fireplace.

The amount of air a fan can move is directly related to the "back pressure" or resistance that it sees. If there were no ducting at all, the fans would do a much better job than they do in real life. The size of duct, the length of the run and the number of elbows all add to the back pressure. In this respect, downdraft units—mounted on islands or peninsulas where the ducts must bend at least once— are inherently less efficient than an updraft unit with a straight duct. For the maximum utility with the least noise it is important that duct runs be kept straight and as short as is feasible. Every bend will mean less air but no less noise.

Ducting and Noise

People don't use their kitchen exhaust systems as much as they should because of the noise they make. Unless the smoke arising from the cooktop is severe, the fans are usually not turned on and, over time, the grease and smells, which could be exhausted outside, end up in drapes, carpets, furniture and on walls. A system that has the least noise for a given amount of air will be the most effective because it will be used more.

There are four things that can help:

- It is a general rule of hydraulic engineering that, for maximum effectiveness, you avoid bends and length in a pipe carrying a fluid. Air is a fluid and *an exhaust system vent pipe should be as short and straight as possible.* When it is not, it takes a bigger fan to get the air through and a bigger fan means more noise and more money.

- Have the fan toward the exhaust end of the vent pipe, getting it away from the kitchen. The attic is a usual location. But it is necessary to have access to the fan for cleaning and maintenance.

- Use a fan design that moves more air with less noise. A "squirrel cage" design does this, but is more expensive.

- Use a larger than standard-sized exhaust pipe. (The noise of air rushing through a pipe is related to the speed of the air and a larger vent pipe means lower air speed.) Making a larger vent pipe will require a modification of the hood itself but may be worth it.

Careful design is imperative for effective operation. To exhaust the same amount of air, a better installation means the fan can be run at a lower speed, and hence lower noise, than in a carelessly installed unit. If you have the chance, you should talk to your designer about this so that he makes special provision for the exhaust vent pipe if that's needed. Then, when the time comes, make sure the subcontractor who is installing the exhaust system knows exactly what you want.

Don't be like the builder who built a neighbor's house. The exhaust fan had been completely ignored during framing and when the time came to install it, the only option was out the back of the house. The fan doesn't work well and the outside flap is noisy. It really does need to be planned for.

A rheostat for controlling the fan speed makes more sense than a simple hi-lo switch arrangement. This gives you the capability to adjust the fan speed to fit the job while keeping the noise to a minimum. Alternately, a three-speed fan is simpler to operate than a rheostat, albeit a little less flexible.

Downdraft Considerations—Because of the larger amount of air they must move, the efficiencies of downdraft exhaust systems are much more sensitive to the length of the exhaust vent and to bends in it. Further, the nature of the system requires at least one bend. Which is why downdraft systems are so noisy in general.

The simplest way would be to run the vent out the back of the cabinet and through a wall to the outside. But on an outside wall you can use an updraft unit that will be less expensive, more effective, and quieter. About the only reason you might to want to use a downdraft unit under those circumstances is if you can't stand the sight of a hood.

If your kitchen has a slab floor, the vent pipe requires special attention. It must be approved to run under the slab using PVC or other plastic pipe. And it must have someplace to go which isn't always easy from under a slab. (All of which is a major reason downdraft units are seldom used under slab floors.)

Updraft Hoods

UPDRAFT FANS WILL BE EFFECTIVE ONLY FOR THE STOVE BURNERS THAT ARE DIRECTLY UNDER THE HOOD OR MICROWAVE OVEN. Fans in microwave ovens are not going to do a good job. You can see this for yourself if you visit someone who has one. Boil some water and turn the fan on. When the pot is on a front burner, most of the steam goes into the room but when it is on a back burner, most of the steam will be captured. (And this happens even though a typical microwave oven fan is rated at 300 CFM, which is about 1.5 times larger than most updraft units.)

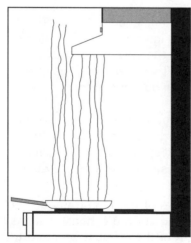

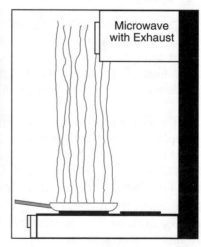

Cooktop and hood are a good combination.

The exhaust in a microwave oven is OK for the back burners.

One way to understand why hoods are effective for only the burners under them is to consider just how fast a fan makes the air move. With a typical fan (200 CFM), the average speed of the air moving across the bottom of a standard hood is 0.6 miles per hour! This is the gentlest of breezes. So it's not surprising that the hood is effective only for things that rise up into it by themselves. And this is why the exhaust fans on microwave ovens are not effective for the front burners of a cooktop or range. Using a 300 CFM fan obviously helps, but 1.5 times "almost nothing" still isn't very much.

Not all updraft hoods are the same size. Smaller ones are used when the cooktop is against a wall and larger units are used for a peninsula or island. (Island cooktops usually use downdraft units, but not always.) When there is no back wall to block stray air movement, the larger hood increases the likelihood that fumes from the cooktop will be captured.

Islands and peninsulas that usually have downdraft systems can be equipped with updraft exhaust systems. There are two ways:

• Mount a large hood similar to those used in restaurant kitchens over the cooktop. This is a very effective system whose main drawback is looks. When it's made of copper and kept clean even the looks are OK, particularly in a country-style house. (Don't forget lights. This type of hood may not have the built-in lights we are used to.)

• Have the hood built into a set of cupboards. By "hiding" the hood in the cabinet, both aesthetics and functionality are served. The photo shows what one builder did using a hood specifically designed for this type of application. Note that this is probably not feasible with a vaulted ceiling.

An updraft system over an island cooktop. Good idea.

Downdraft Options

There are two kinds of downdraft systems: one where the exhaust intake is on the same level as the burners, usually between them, and one where the exhaust intake is across the back of the cooktop. This may be fixed in place or it may be raised up and down as needed.

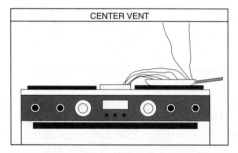
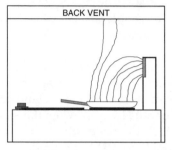

There are two types of downdraft systems.

Ambient Air Movement

It was enlightening to actually watch an exhaust fan do its job when barbecuing on an island grill with a center-vent downdraft unit. When the air in the room was quiet, the fan got rid of most of the smoke, but not all. However, there was a ceiling fan about 12 feet away in the kitchen nook and, when this fan was on, the effectiveness of the downdraft fan was clearly very poor.

A similar test was carried out with an updraft fan over a peninsula cooktop. The hood apparently protected the cooktop to a large extent from the ceiling fan. There wasn't a noticeable difference when the ceiling fan was on or off.

Downdraft systems with the exhaust in the center between burners will be even less effective in venting steam from a tall pot of boiling water, such as is used to cook pasta. The back vent units will work better on the back burners, but the ability of the fan to pull steam and smoke sideways from the space over the cooktop will always be a limitation—particularly for the front burners.

Ductless Exhaust

"Ductless," "recirculating," "no vent," or "no duct" is a nice way of saying that whatever the fan picks up from the stove will be passed through a charcoal filter and dumped back into the room. When it's not vented outside, the exhaust system does very little good. While the heated air, smoke, fumes, and steam coming up from the stove are run through the filter, it removes only a little of the smoke and fumes and none of the water vapor or hot air. In short, it is about as close to useless as you're going to find. A maga-

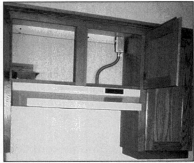

These have no obvious vents. If there is one out the back of the house, fine. But don't get one that vents back into the kitchen.

zine article rated this arrangement as "five percent effective."

The usual place where you'll see a ductless or re-circulating exhaust system is in a microwave oven. But, microwave ovens aren't the only place. National manufacturers of kitchen exhaust systems make updraft units that aren't a part of a microwave oven but which are intended to be used in a ductless mode.

Finding these ductless systems is easy. If the power is on, simply turn the fan on and see if the air comes back into the room from the louvered area at the top of the microwave or hood. If it does, you're looking at a ductless exhaust. If the power isn't on, open up the cupboard above the microwave to see if there's duct work there. If there is, fine. If not, you've probably got a ductless exhaust, although it is possible to vent the exhaust directly through the back wall without any duct work showing in the cabinet above the unit. To be sure, check it with the power turned on—or look outside where the vent should be.

Ductless systems are also available for downdraft units. They won't be effective there, either.

One of the most annoying things about these systems, aside from their uselessness, is that they often blow the air back into the room at just about head height. Having a blast of whatever is coming from the stove blowing in your face is not very friendly.

O ne woman reported that, when she remodeled her house, she had a microwave oven with a ductless exhaust fan installed. With the combination of air blowing in her face and the overall uselessness of the unit, she opens a window next to the stove when she needs to have smoke exhausted. Most of us don't have windows close to the cooktop, nor do we feel like pouring money down a rat hole to get a useless piece of machinery.

Another housewife's comment was, "I hate it." She is saddled with a ductless exhaust in a house bought when she didn't know better.

Manufacturers don't recommend the use of ductless units if there is any way to duct them outside. But, as one customer representative said on the telephone, "They're better than nothing." Well— maybe. They're used in new houses, not because there's no other choice, but so the builder can save a few dollars. They add glitz by paying attention to customer's expectations but there's no substance. They can get into new houses for one or more of these reasons:

1. The builder doesn't know better.
2. The builder is trying to save a few bucks on the duct work.
3. The builder is an area where the practice is to use the ductless fan when the hood isn't on an outside wall and he's following local custom.

In any case, the use of these ductless arrangements is about as user-hostile as you get.

Other Exhaust System Considerations

The height of an exhaust system above the cooktop is important. The lower the exhaust system, the better. But too low will put the hood or microwave oven in the way for cooking. Too high will make the exhaust system even more sensitive to extraneous air movement. A suggested compromise is 24 inches for hoods that stick out in front and 21 inches for microwave ovens. (24 inches would put the bottom of a microwave oven just five feet above the floor, making it difficult for a short person to see inside.)

A common complaint about exhaust hoods is the appearance. There are two designs that address this. Both use movable parts that are brought into use only when needed. One is a simple flap that is pulled out from the front of the hood to try to increase the effective size of the hood.

In the other design, the part that pulls out is much more than just a flap. It is the full depth of the hood and includes a part of the exhaust system filter. When it is not in use, the movable part is back even with the front of the adjoining cabinets. The front of the hood may be glass, black or white enamel to match the stove, or the same wood used in the cabinets. The unappealing appearance associated with hoods is gone.

The entire under surface of the hood is covered with filters. Lights, which are an integral part of the hoods, are above the filters so that they must shine through to illuminate the stove top. The manufacturer (Imperial Cal, Irvine, California) offers these in versions with 300 and 400 CFM, both of which are larger, noisier, and more effective than most updraft exhaust systems.

Keeping It One Way

With an exhaust system, a flap in the duct prevents outside air from coming into the house when the fan is off.

With a vertical exhaust vent, the flap is normally in the hood, right in the kitchen. If this flap sticks open, it lets cold air into the kitchen and makes a startling noise when it finally closes. If possible, check the action of this flap before the hood is installed. It should firmly return to the closed position when the fan is turned off. If you run into the problem while the first year's warranty is in effect, the builder should fix it—but good luck.

We have this problem in our house. We try to remember to check the flap by putting a hand under the fan in the hood. If we feel cold air, a solid thump on the hood will usually close it. The flap sticks open when the exhaust fan has been run or when a strong wind causes suction up the pipe. Obviously it is better, if you can, to avoid getting a hood with a temperamental flap.

With a horizontal vent pipe, the flap is outside the house at the end of the pipe to keep rain and cold air out. This flap rattles and can be quite noisy when the wind blows. For this reason, don't use such an arrangement if there is any reasonable way to have a vertical vent. Unfortunately, for downdraft systems you don't have a choice.

Chapter 18

Other Kitchen Considerations

Breakfast Nook and Family Room

In today's open floor plans the family room and kitchen are often more like one room than two.

You might come across a potential floor covering problem when a peninsula or island eating bar separates the kitchen and family room. Many builders put the family room carpeting all the way to

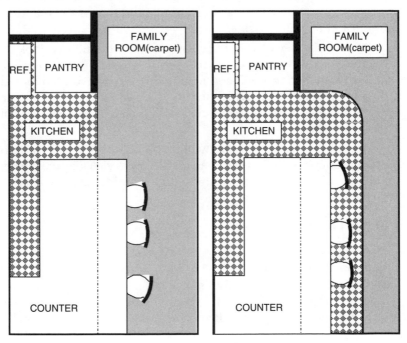

Carpeting doesn't belong under the breakfast bar.

the bar. Food and drink spills can ruin your carpet, so you may want to consider extending the kitchen floor covering in that area.

The Breakfast Nook

Is the breakfast nook large enough for table, chairs, and still have room to walk behind the chairs? For room to get behind people seated at a table, allow 36 inches between the edge of the table and the wall. And watch out for hinged doors as discussed in Chapter 6.

If there is a kitchen eating bar, is it convenient? If such a bar represents the only eating area in the kitchen, it might be well to reconsider the whole design. Most folks prefer a sit-down table for family meals.

(Also *see* the suggestions regarding lighting in the breakfast nook later in this chapter.)

Stoves, Cooktops and Ovens

The configuration of the stove's cooktop and oven or ovens are decided as a part of the house design. You will need to decide if the stove is to operate with gas or electricity and then select manufacturers and model numbers.

Be careful when comparing houses that you don't overlook the microwave oven. In an effort to keep the basic price of a model down, many builders will not include the microwave even though they know that buyers will want one. The worst part of this practice is that they also leave out a reasonable place to put it when you move in.

And putting a microwave oven on a counter is a real waste of often-precious space. It's not a bad idea to try to visualize the kitchen with the microwave oven in it and see if the builder will accomodate you and how much it's going to cost.

Oven Door Clearance

There should be 48 inches clearance in front of a thermal oven so someone can comfortably stand in front of the open door and take out a roast pan. Most oven doors stick out 24 inches. A person leaning over to take something out of the oven needs at least another 24 inches or a total of 48 inches. If you are going to be using the smaller ovens with 21 inches doors, you can cut the 48 inches to 45 inches. When you don't have this clearance, you have to take pans out of

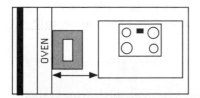

the oven from the side which puts you in danger of burning yourself. Caution: this argument applies to ovens that are a part of a stove as well as to cabinet-mounted units.

Not enough room can be dangerous.

Oven Height

Be sure to specify how high you want the ovens, both thermal and microwave, above the floor because these dimensions vary widely from one builder to another. The height of thermal ovens should not make it difficult for a short person to get a hot roasting pan out of an oven.

This seems obvious, yet I have seen ovens so high that it would take a Michael Jordan to use them comfortably. In one case it was in a custom house and I can only assume that the new owner was a very tall person. In another case it was in a spec house and it was hard to understand the logic.

If you are tall and want an oven way up there, be well aware that when it comes to resale time you will really restrict the number of potential buyers. Better you should lean over a little bit to see in the oven and not seriously lower the value of your property.

For two reasons the best design is to not use a microwave oven installed over a cooktop:

- It will always be too high for short homemakers who cannot see into it and cannot safely handle hot dishes that are over shoulder height, and

- As discussed in the previous chapter, these ovens make second-rate exhaust systems.

If you really must have the microwave over the burners, then put the it as low as you can and still have room to work on the cooktop. Placing the bottom of the microwave oven about 21 inches above the cooktop is about as low as you should go even though this is still too high for short people.

When the bottom of a wall-mounted microwave oven is in the order of 45 inches above the floor it is convenient for everyone.

The Pantry

Both walk-in and cabinet pantries are seen in today's houses. There are advantages and disadvantages to each.

Walk-in pantries should be convenient to the kitchen working area. Take a hard look at any walk-in pantries to figure out just how useful they will be. Pay particular attention to how much useful shelf space you have. Also, make sure that when you open the pantry door, you're not blocking access to shelves.

With walk-in pantries, be careful of where the light switch is placed. Outside, *next to the latch side of the door,* is fine. As with other light-switch locations, the thing to be careful of is to not put the switch where (1) it is not expected or (2) behind a door. And don't put it inside the pantry hidden under a shelf. Hence the suggestion that it is better to have the switch outside than to have it inside in a bad place.

And don't forget to check that there is a light in the pantry. In one set of tract models there was no light and no kitchen light that shone in. You wouldn't usually notice it in daylight but after you moved in you'd need a flashlight to find things.

Cabinet-type pantries are usually a continuation of the cabinets in the kitchen and have the same depth as the under-counter cabinets, that is, about two feet. There are two distinctively different types of pantries that are made as a part of the kitchen cabinets: those with horizontal shelves and those with tall hinged vertical panels.

Horizontal shelves should be pull out, otherwise the back half won't be useful unless you enjoy unloading the front part of the shelf to get to something in the back. In the other type of cabinet pantry, narrow shelves are mounted on the back of the front door, on both sides of hinged vertical panels and across the back of the pantry. Access is by opening the front door and swinging the vertical panel to readily reach any stored item. Because of their cost, don't expect to see these in tract models. You will pay extra to get them.

While they may be okay with you, the wire shelving seen in numerous tract houses in all price ranges just doesn't make sense to me. Small bottles and packages should sit level which they won't do on the wire. Wood does a better job.

Power and Light

General Kitchen Lighting

If your kitchen has a flat ceiling there are several useful lighting options. This is a place where regional differences are noticeable; fluorescent panels are popular in some areas while floodlights are usual in others. Track lighting can also be used but many people object to the aesthetics.

If you are looking at a vaulted ceiling in the kitchen, remember that it will restrict the kind of general lighting you can use. Fluorescent panels, for example, won't do as good a job when mounted on a sloping ceiling. And recessed lighting systems must be designed carefully to make sure the lights point downward and that the coverage is uniform.

If you have the option, choose what you want but make sure you have enough light in the kitchen.

An attractive scheme was seen in the kitchen of one upscale house that has a large decorative translucent panel with a skylight above it in the center of room. After sundown fluorescent lights mounted in the well of the skylight are used to continue the effects of the decorative panel.

Light for the Cooktop

If you use an island downdraft exhaust system, check that there is good lighting for the cooktop because there may be no special light for the island unless you ask for it. Sometimes a skylight is used to provide light during the day, but night can be another problem.

Light for the Sink

Most houses have a window in front of the sink. Large windows make the kitchen feel more open but take away wall space that might be better used for cabinets. It's a matter or personal preference.

Just when you think you've seen it all, along comes another one. This was in an above-average-price spec house being shown as a part of a tour. The sink was on the side of a U-shaped kitchen but not facing a window. And there was no light over or around it! Whoever bought that house has a problem.

Light at night for the sink work area is usually as simple as installing a light when the house is built. But sometimes it is forgotten. One attractive, but different, arrangement has been seen in a few model homes. It includes a problem you'll want to be careful of.

The sink is in front of a tall window that goes to the ceiling. The ceiling has several feet of glass that comes down to the tops of the windows, giving a solarium effect. During the day, all that light and airiness is appealing. But at night there is no provision for lighting the sink and surrounding countertop!

If you run across this arrangement in a house design, be sure you figure out a way to get light around the sink or forget the design. If you decide to use it anyway, don't forget to check the orientation of the kitchen relative to the sun, because, since there is no obvious way to put shades on these windows, it will be a problem for south or west facing kitchens.

Counter Lighting

Cabinets often shade portions of the counter work areas from the regular kitchen lighting. While small fluorescent fixtures can be hung under the cabinets at any time, it is better to have them wired in place so you don't have cords running around to the nearest outlet. This lighting is another inexpensive functional feature that will make working in the kitchen easier.

The Breakfast Nook

Light in the breakfast nook is usually provided by a small chandelier or some kind of hanging fixture. If you prefer a ceiling fan in place of the chandelier or may want to change it later, be careful. When the chandelier is installed, have its mounting made strong enough for a ceiling fan. Have a 3-wire cable run to the switch location. Depending on the exact layout of the nook, be careful about the placement of the overhead fixture. If you're not, you may find that the chandelier isn't centered over the table and it may not be that easy to change. Chandeliers can be swagged but ceiling fans present more of a problem to move.

(A "swagged" ceiling fixture is one where the electrical box is mounted in one place and the fixture hangs from another. A loop of wire and chain, the swag, is then necessary between the electrical box and the top of the chain holding the fixture.)

Some builders solve the question of where to put the chandelier by lighting the nook with can lights. This is okay as long as no one ever wants a ceiling fan. Just make sure there are enough lights (which builders don't always do).

Instant Hot Water

Depending on the layout of the house, the kitchen sink may be many feet away from the hot water tank so that you run gallons and gallons of water down the drain several times a day before hot water reaches the kitchen from the water heater. You may decide that in these days of water shortages it would be not only convenient but also socially responsible to have an instant hot water unit.

These well-insulated units hold about a gallon of water and the water is kept hot electrically. A separate faucet on the kitchen sink gives you access to hot water instantly.

Be sure there is an outlet under the sink where it can be plugged in and be sure the available socket is on all the time and is not switched on and off by the disposal switch.

Outlet Under the Kitchen Sink

Where the builder is trying to save pennies, the practice is to not put any outlet under the sink at all. The disposal and the dishwasher are permanently wired in place. If left to the electrician, this may be what happens because it's the least expensive. But if any work ever has to be done on one of them or if either must be replaced, it requires both an electrician and a plumber. Thinking ahead will come up with a better scheme.

A better practice is to have a split-wired dual outlet in the back wall under the kitchen sink. One side of the outlet is wired to the garbage disposal's on-off switch and the other side of the outlet has power all the time. There are two ways this outlet is used:

• Sometimes the dishwasher is wired permanently to the house wiring which leaves the hot half of the outlet available for an instant hot water unit.

• In other regions the dishwasher plugs into this "hot" side of the outlet. This is the preferable way. It means the electrician is not needed when work has to be done on the dishwasher. But it leaves no place to get power for an instant hot water unit—unless a second outlet is available.

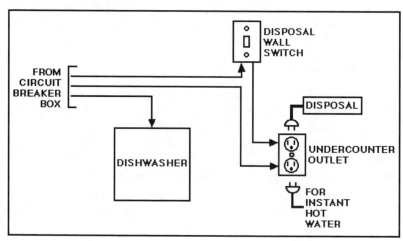

Arrangement for instant hot water using one outlet.

So when the wiring is done, try to get two outlets.

One is split for the dishwasher and the disposal—the disposal half is switched. The second outlet is for instant hot water if it's

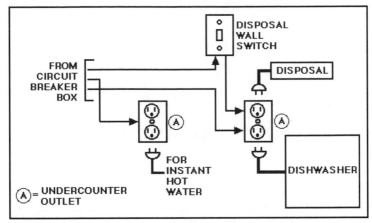

With two outlets, the dishwasher can plug-in.

ever wanted. Note that the cost of putting this outlet in place initially is low. To have it done later would probably be prohibitive.

(Don't plan to use the line to the dishwasher as the source of power for the instant hot water. The dishwasher takes 15 amperes and is on a 15- or 20-ampere circuit. If you add the instant hot water unit (about 6 amperes), you can overload the circuit.)

Outlets in Islands and Peninsulas

The National Electrical Code requires that there be at least one outlet in an island or peninsula. Careless, unthinking electricians will sometimes put these in a beautiful hardwood cabinet where they stick out like a sore thumb. Insist on having a say in precisely where any outlets are placed. It's not a bad idea to wait until after the cabinets are installed before you make the final decision. Alternatively, consider having the outlets completely hidden by mounting them face downward under the overhang of the countertop. (You must plan ahead for this, however.)

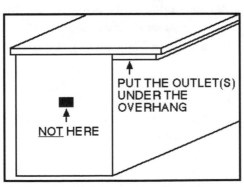

Outlets shouldn't be an afterthought.

Chapter 19

Kitchen Layouts

Whether working with plans or looking at models, it is very useful to recognize the basic kitchen designs and the pros and cons of each. It makes it easier to decide how well a particular arrangement will work out and whether it is appropriate to the house you're considering.

Within the almost infinite variety of kitchens, there are a relatively small number of basic designs in common use. Their advantages and disadvantages are discussed here along with some precautions that should be taken with each.

Kitchen Design Basics

Other than how it fits into the whole house, there are only a few basic constraints on good kitchen design. These common-sense considerations are:

- Dishwashers should be close to the sink, close to cupboards, never in a corner, and not in the way when the door is open.
- Refrigerators should be located where all doors can be fully opened. (No wall next to a refrigerator door hinge.)
- The kitchen should not be a traffic path from one part of the house to another.
- The *walking* distance between the sink and the cooktop and between the sink and the refrigerator should be kept short.
- There should be landing areas—places to set things—convenient to the cooktop, wall ovens, refrigerators, and cabinet-type pantries.

Overall Considerations

Work Areas

Most kitchen work is limited to a few areas, the primary one being around the kitchen sink. Secondary areas include the cooktop, the ovens, and the refrigerator. Note that the dishwasher is not mentioned separately because it is (or should be) next to the sink.

The Primary Area

You need counter space for preparing foods, you need a place to soak bowls and pans after they're used, and you need water handy for rinsing your hands. After the meal you need a place to set dirty dishes before they're cleaned for the dishwasher. Watch in your own kitchen and you'll find that it is the area around the sink where most of the work is done.

You need at least a couple of feet of counter space on both sides of the sink. And you need plenty of light. Many kitchen layouts have the sink in front of a window both for light and to break the monotony of a bare wall. And the larger the window the better—except for one consideration. Upper cabinet space close to the dishwasher is usually at a premium and you can't put cabinets where you have a window. Since the dishwasher should be next to the sink, modest window sizes are appropriate.

Secondary Areas

The counter next to the cooktop should be open for two reasons: you need a place to set a hot pan off of the burners and you don't want grease spattering on an area that is hard to clean.

I know—in this day of fat consciousness you don't fry or saute as much as in the past, but it's still better when cleanup is easy.

You need a place to set things going into or coming out of the refrigerator and the oven. These landing areas are usually present in a kitchen design but occasionally you'll see a problem, particularly with refrigerators, as discussed in Chapter 15.

Be sure there are work areas around the ovens and the cooktop and there are landing areas for the refrigerator and the ovens.

The Kitchen Triangle

Kitchen designers have made a big thing of the kitchen triangle—the triangle formed by the sink, the refrigerator, and the stove. The National Kitchen and Bath Association sets the guidelines for the maximum size of this triangle at 23 feet. In book racks at home-improvement stores you'll find a bundle of how-to books, many of them dealing with how to design kitchens and bathrooms. Watch out! These books have little to do with a user-friendly home; the designs are not for the convenience of the people who will be living in them—the stress is on how it looks, not how it works. Remember? It's the pretty pictures that sell these magazines and books, not the quality of the designs.

The stress on the triangle is hard to understand. To be sure, the number of steps from the sink work area to the refrigerator is important and so is the spacing between sink and cooktop. But the distance from the refrigerator to the cooktop is almost trivial because so seldom do foods go directly from the refrigerator to the stove without a stopover at the sink work area—for unwrapping, if nothing else. (This isn't the case in restaurants where large amounts of food are prepared for the stove and stored in the refrigerator ahead of meal time and, when an order comes, it *does* go directly from the refrigerator to the grill. But a restaurant is not the same as the family home.)

It is recommended that you concentrate on the distances between the sink and the cooktop and between the sink and the refrigerator and don't worry about how far the refrigerator is from the cook top. The greater the distances, the more steps it makes every time you use the kitchen, so keep them short. Measure off for yourself the difference it would make whether you have to go, say, 4 feet from the sink to the cooktop against what it's like if they're 6 feet apart. Remember these are distances you walk and, if there's an island in the way, that you have to go around it.

Appliance Clearances

Don't forget to have adequate space in front of appliances; 42 inches minimum for dishwashers, 48 inches for ovens, and 64 inches from the back wall for refrigerators. (If the available space won't permit these minimums, smaller spacings can be used with a concomitant loss in functionality.)

And be careful of two appliances opposite each other where you can't get through when both doors are open at the same time.

Corner Sink

When the kitchen is on a corner of the house, it can double the amount of window area in front of the sink without separating the cabinets from the dishwasher. It is useful for all but the in-line and the galley layouts. The cautions about dishwasher location were given in Chapter 14.

Basic Layouts

Virtually every kitchen is unique, yet they are all some form of one of these basic layouts:

- In-line
- Galley or corridor
- U-shaped
- G-shaped
- L-shaped
- Box

In the following discussions take a look at how each meets the criteria listed on the first page of this chapter—some do a better job than others.

Because of the relative complexity of the kitchen, be sure it's a high priority consideration in deciding which overall design you select for your new home.

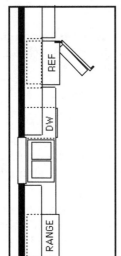

In-Line, The Simplest of All

Although almost never done, there is nothing wrong with putting all appliances and counters in a straight line with the sink between the range and the refrigerator. Since there's no landing area in front, only top-mount or bottom-mount refrigerators should be used (a drawback for someone wanting a side-by-side).

The primary design consideration is to keep the range and the refrigerator as close to the sink as possible (to minimize steps) while at the same time leaving adequate work areas on both sides.

The in-line kitchen.

The area in front of the counter should not be a major traffic arterial. This can be a serious problem in a small house and is one of the potential disadvantages of the in-line layout—the other being the scarcity of walls for cabinets.

And You Thought a Galley Was on a Ship!

The galley or corridor is another efficient kitchen design, efficient in that most of the work can be done with very few steps—everything is right at your fingertips. But it is tight—there is no place for two people in this kitchen.

The nature of the layout lends itself to being a traffic aisle which is one of the reasons it is not used more often. And in today's kitchens we like openness which this kitchen does not have. It has a closed-in feeling and doesn't share its space with an eating area or the family room.

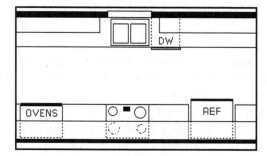

The galley kitchen.

The layout precautions are not unusual. Be sure there is enough space between the counters for open appliance doors. (48 inches is recommended.)

The U-Shaped Kitchen

The U-shaped is probably the most common of the different kitchen configurations. We subdivide it into three arrangements—the narrow kitchen with the dishwasher on one side, the narrow kitchen with the dishwasher at the end, and the wide U-shaped kitchen.

When one end of the galley kitchen is closed, we get the narrow U-shaped kitchen with the sink on one side. This, too, is an efficient arrangement but it does have two cabinet corners that need lazy Susans to make them accessible.

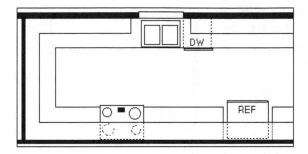

The U-shaped kitchen with the sink on one side is a good design.

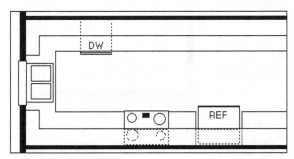

Narrow U-shaped kitchens should not have the sink at the end.

When the sink is put at one end of the narrow U we have an arrangement that is *not* satisfactory because there is no suitable place for the dishwasher that doesn't separate it from the sink or doesn't put it in a corner.

When the U is made wider, there is more flexibility in sink and dishwasher placement and there is room for more than one person to work in the kitchen at a time. It is often used with a central island. The open end of the U is also a good place for the kitchen table if there is no separate nook.

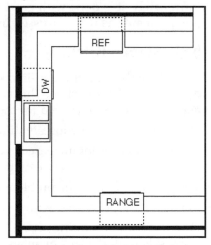

The center of a wide U-shaped kitchen is often the eating area.

The G-Shaped Kitchen

When a wide U-shaped kit–chen has a peninsula added, partially closing off the open side, we have the G-shaped kitchen. The peninsula offers an excellent site for a breakfast bar. This lay-out is flexible, reasonably efficient, and very popular.

With the G shape there are now three corners in the cabinetry. The one at the peninsula can be accessed through a door from the back side (*see* arrow in the diagram), making it a useful place to put things that are not used every day. But do be careful; this door doesn't **have** to look as if it were an afterthought, even though it usually does.

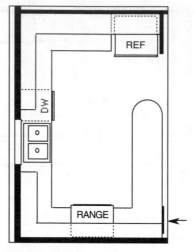

The end of the peninsula makes a good landing area for the refrigerator so that side-by-side as well as freezer-over models can be used.

The G-shaped kitchen is popular. It is an efficient design when used correctly.

The peninsula is sometimes used as a location for the cooktop. The same safety concerns apply here as on an island.

There is one particularly bad way to use this arrangement and, unfortunately, it is done with great regularity. *Don't* have the sink in the peninsula as was discussed in Chapter 14.

The L-Shaped Kitchen

If we take the end of the in-line kitchen and bend it 90 degrees we get an L-shaped kitchen. This is often seen in country kitchens where the open space is used for the family eating area.

Three things are functionally different from the in-line model:

• There is a significant increase in counter space without much increase in the distance between the range and the sink.

• The potential of a traffic problem has been greatly diminished,

• There is a cabinet corner with its inherent inconvenience.

The refrigerator landing area is still the adjacent counter. For a side-by-side unit, the lack of an area in front will make it inconvenient to use—which leads to the desirability of adding an island to this layout.

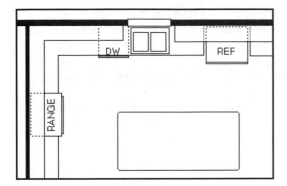

Islands add significantly to the utility of L-shaped kitchens.

Islands

Islands can be added to the L-shaped, the wide U-shaped, the G-shaped, or the box layouts. (The box layout is described below.) When we put an island in a kitchen, we take away the room for the family table but add a significant amount of working counter area as well as cabinets under the island.

There are several ways this island area can be used—and mis-used:

•**Don't** make that most common energy-consuming error—your *personal* energy that is—don't put the refrigerator across the room from the sink on the far side of the island. Islands are great but not between sink and cooktop or sink and refrigerator.

> 'll always remember a house I saw in a special "Showcase of Homes" sponsored by a prominent building designer. The house itself was his own new home and one assumes that it represented his best effort. It had a U-shaped kitchen with a large island and...you guessed it: the refrigerator was on the far side of the island from the sink work area. It will always be a reminder to me about how insensitive designers and architects can be to the functional aspects of their designs.

• A real plus for an island is that it automatically becomes a landing area allowing you to efficiently use any type of refrigerator.

• When the island abuts the breakfast nook or the family room, one side of the island can be—and often is—used as an eating bar.

• If the cooktop is there and the island is to be used as an eating bar, be sure the bar is raised several inches *higher* than the cooktop to minimize the potential for burns from burners, upsets, or spattering food stuffs. Kitchens where the bar is lower than the cooktop are disasters-waiting-to-happen. Not only is there the danger from spattering grease, now scalding water from pots and tea kettles is an ever-present danger to someone at the eating bar.

I'm a safety nut and when I see an eating bar and a cooktop on the same counter it waves a red flag at me. The potential for injury, particularly to children, is too great.

- If the cooktop is in the island, you will have to use either a downdraft exhaust system or, preferably, a ceiling-mounted updraft hood, either of which will add cost to the kitchen.

- Leave enough room in front of the appliances to be able to open their doors without causing problems. The diagram on page 201 is scaled for our recommended 48 inches between the counters and the island.

- If the sink and dishwasher are in the island, have the dishwasher on the side of the sink closest to the majority of the upper cabinets. This puts the open dishwasher door near the cabinets, making an efficient arrangement.

- Have the shelves in the island base cabinets accessible from both sides so that things towards the backs of shelves will always be accessible. Pull-outs are usually a good idea.

An L-shaped island kitchen. Note the cabinets over the peninsula and the space beside the dishwasher. Very well done!

The kitchen in the photo is an example of how simple common sense can be applied to design. Functional and beautiful. This was not done by a professional designer but by the wife of an experienced builder. She and I are fully in agreement that most kitchen designs need a woman's touch and insight—regardless of the designer's gender.

The Box Layout

When you start thumbing through a book of home plans or looking at models, it won't take long to find a kitchen that has counter space and/or cabinets on all four sides. This is the box kitchen. It has at least two doorways, sometimes three or even four. The usual arrangement is an extension of the wide U-shape with an island and, across the far side of the room, cabinets, a desk, and/or the pantry.

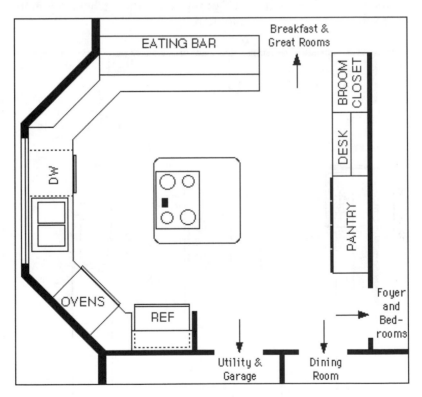

A box kitchen is the cornerstone of many large homes.

The Box Kitchen Turned Bad

A sure sign that neither the builder nor the designer ever worked in a kitchen is to find layouts where the refrigerator or ovens are completely across the room from the sink's working area, often on the far side of an island. This seems to happen more in upscale houses, perhaps because the designer and builder are so busy making beautiful houses that they never involve themselves with food preparation or cleanup. Here is the essence of one kitchen that sticks out particularly in my mind. It epitomizes the problem.

Putting the refrigerator on the far side of the room, many steps away from the food preparation center, is inexcusable. Sure it makes an aesthetically "clean" design—those odds and ends, the pantry, the refrigerator, and the desk, are put out of the way where they don't interfere with the "feel" of the rest of the kitchen. Remember, "form follows function?" Not here.

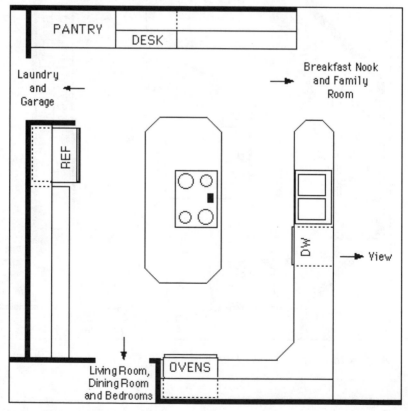

A box kitchen you don't want.

Perhaps even worse than putting the refrigerator on the far side of the island is to put the stove there. Watch for that too and don't have it in your kitchen.

Another thing that this kitchen has, that you surely won't want in yours, is the dishwasher-on-the-peninsula problem discussed earlier—there are absolutely no cabinets for dishes that aren't several steps away from the dishwasher. Again a case of aesthetics before function. The view from this house is spectacular, looking out across the rolling hills of the Willamette Valley with its fir, pine, cedar, maple, and oak. And whoever works in this kitchen can enjoy that— even while being unhappy about the functional arrangement.

Look at this kitchen's relationship to the rest of the house. Anyone coming from the bedroom wing of the house and going to either the family room or the garage will walk through the kitchen. Laundry will be carried through it. Even though this traffic is not through the work area between the sink and the cooktop, the kitchen should not be like the lobby of a public building.

At least in this design there are good clearances in front of the appliances and neither the dishwasher nor the refrigerator is in a corner. And there *is* a nice landing area for that lonesome refrigerator.

Part IV

Beyond the Kitchen

With the subject of the kitchen done, this part of the book looks at the rest of the rooms in the house. In Chapter 20 attention is paid to potential problems in bedrooms and baths and how avoid them.

The formal dining room is singled out because it is so often designed with little thought to what will be in it. And in Chapter 22 the laundry is not overlooked because it, too, needs more attention than it usually gets.

The garage can be a blessing or a curse and should not be forgotten when looking at how houses are laid out.

There are a number of tips in Chapter 23 on the little items outside the house that make a big difference after you've moved in.

Chapter 20

Bedrooms and Baths

Bedrooms and bathrooms require critical decisions about lighting, furniture placement, windows, vanities, tubs, showers, and switch and outlet locations. Wherever you can make the decisions, your new home will probably be the better for it.

Bedrooms

Lighting

There are five different arrangements for lights in bedrooms.

- No fixture. The power to an outlet is controlled by the room light switch and the light is provided by a plug-in lamp. This is the least expensive and is the usual set-up in tracts where every dollar counts.

- Room-center fixture. A ceiling fixture, usually small, is located in the center of the room. It is controlled by the room light switch. This is a common arrangement outside of tracts.

- Recessed can lights in front of closets. One or more lights are placed in the ceiling to illuminate the room and particularly the closet. This is also common outside of tracts.

- A ceiling fan is provided with a light kit that is controlled by a wall switch. This is used less commonly.

- Indirect ceiling lighting is sometimes used in large bedrooms with high ceilings to give a dramatic effect.

If the room-center fixture might be changed to a ceiling fan in the future, the mounting box should be strong enough to support the fan. A 3-wire conductor should be used between the fixture and the wall switch location. (*See* Chapter 5.)

Where you have choice, you'll need to decide what you want in each of the bedrooms. The recommendation is to go along with local custom unless you have a reason to use a different option.

Don't forget to check the closets in each bedroom. Walk-in closet doors should not block clothes when they're open and the door openings in other closets should not limit access (*See* Chapter 12). A common problem is for the bedroom door to swing across a closet door. You should never have to close one door to open another. (*See* Chapter 9). See that the closet doors are the style you want.

The Master Bedroom

Furniture Placement—The master bedroom, like the dining room, is often designed without any thought for placing furniture along the walls. A typical master bedroom suite consists of the bed, two night stands, a large dresser, and a chest or armoire. Besides these, there is often a TV stand or a wall unit for holding the TV, books, etc. When the designer includes several windows and unusual wall angles, there may be simply no room for the dresser—to say nothing of the chest or armoire.

In some larger homes, cabinets will be built-in in the walk-in closet, obviating the need for room for a dresser and a chest or armoire. Or the chest or armoire, at least, are put into the closet, not because that's the most desirable, but because it's the only place available.

When I walk into a model and see the bed at an angle across a corner of the room I always check to see if this is simply a dramatic decorating arrangement or if it's because there's no place else to put the bed and nightstands. It happens both ways. Of course, if you like this and there's room, that's fine, except don't unwittingly get caught with an arrangement you can't change—in case you decide you want to later.

Doors—Double entrance doors have the same potential problems in the master bedroom as elsewhere: they can block light switches and take up useful wall space, leaving no room for furniture. In some designs the walk-in closet door swings into the bedroom rather than the closet. This generally takes away from valuable wall space in the bedroom. A pocket door should be used instead.

Windows—Large windows on the south and west sides of the house can be sun problems in the summer; if possible, face any large windows east or north. Windows, of course, should not be in walls that are needed for furniture.

TV and Phone Outlet—Pick the TV outlet placement carefully. Usually, there is only one wall on which a large dresser can be placed. The TV outlet shouldn't be on this wall unless you want it sitting on the dresser. You'll probably want the telephone jack around the head of the bed so the telephone itself can sit on one of the night stands.

Electrical Outlets—Be sure that outlets will be accessible when you put furniture in the room. (*See* Chapter 5.)

Cabinets—If there is a linen closet in the master bedroom suite that is more than about 12 inches deep, pull-out shelves will make it more useful.

Consider having pull-out shelves in vanity base cabinets. The bottoms of these cabinets are always close to 24 inches deep and the items at the backs are virtually inaccessible without pull-outs. Even if the pull-outs mount as close to the bottom of the cabinet as is possible you lose a little vertical height but gain considerably in overall utility.

Vanities

Lighting

Light bars and fluorescent lights are frequently used to illuminate a vanity area. They cause fewer shadows when applying makeup or shaving. You may prefer the skin tones brought out by incandescent lamps in light bars even though fluorescents use less electricity and generate less heat.

Windows and skylights are also used to provide lighting. The need for concern about excessive summer solar heat through glass was discussed in Chapter 10. It is a particular problem in small areas like bathrooms where skylights are often used to provide outside light.

Counter Heights

Counter heights for kitchens are pretty standard at around 36 inches. In bathrooms that may be used by children, it's common to put the counters lower, typically 32 inches. In the master bedroom vanity area, however, there are a variety of heights, the argument being that the higher counters are more comfortable for an adult and there's no reason to make them low enough for children. You may want to give this some thought.

Countertop Materials

The same materials found in kitchen countertops are also used in bathroom vanities: laminate, tile, stone, solid surface, and solid-surface veneer. Solid-surface counters are seldom used because the additional cost is not warranted for bathroom use. The relatively new solid-surfacing veneer is an excellent candidate for vanities and bathrooms.

Cultured marble is another countertop material that is excellent and popular for vanities. It is not used in kitchens because of its relatively soft surface and higher cost than laminates.

The other materials were described in Chapter 13. We look at cultured marble here.

"Cultured marble" is not what its name implies. It is a cast polymer which can have a marble-like appearance. It is manufactured by molding a mixture of finely ground lime-stone and polyester resin. Ma-

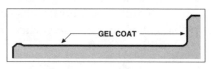

Cultured-marble countertop.

terials other than limestone are sometimes used to get "cultured onyx" and "cultured granite." The resulting products are different in appearance and are a little more expensive then cultured marble.

Cultured marble may be a single color but it can be made with a marbleized appearance that many people find elegant. A well-done cultured marble shower or tub surround, for example, can be quite handsome.

Cultured-marble countertops are usually made with the counter, backsplash, and the front face all molded in one piece. A small lip on the front of the countertop keeps water from running onto the floor. The ends are separate pieces that are installed at the same time as the top. The material is quite heavy and there is a limit to the practical size of a single piece of cultured marble.

The top surface is a gel coating that is about one-fiftieth of an inch thick and has a hardness that is between laminate and tile. Minor scratches in surfaces can be polished out by someone experienced in handling the material. Major damage to gel-coated cultured marble can also be repaired but the know-how to do this is not widespread.

When seams are required, the two pieces cannot directly abut one another because there will be inevitable small differences in thickness that cannot be smoothed out because the gel coat surface is so thin. Thus seams always involve a mastic that is usually quite evident. For the same reason, gel-coated material isn't usually seamed directly with other materials.

With cultured marble, basins can be molded as one piece with the countertop making a smooth unified appearance. This isn't done with other vanity counter materials. A disadvantage to this arrangement is that any problems in the basin will require that the whole countertop be replaced.

The International Cast Polymer Association (ICPA) is the trade organization for the cast polymer industry—including manufacturers of solid surface materials. A serious past problem with cultured marble was the inconsistency in quality and performance. In poorly made cultured marble, particularly for applications subject to hot water such as basins, tubs, and shower pans, the gel coat may crack or craze over a period of months or years. This gave the product a bad reputation that hurt the industr. The ICPA and the National Association of Home Builders now have a certification program to ensure that products with the NAHB-RC/ICPA certification label meet the high quality that should be present in a good product. This certification label should be present on all cultured marble made by ICPA members. Look for it; it is your assurance that you are getting an approved product.

Costs

Laminate is the least expensive countertop and solid-surface the costliest. Tile, stone, solid-surfacing veneer, and cultured marble are in between.

Basins

Basins may be dropped in, integral with (in the case of cultured marble), or mounted under the countertop.

Drop-in Basins—There is the same objection to drop-in basins in the bathroom as in the kitchen: the basin lip sitting on top of the counter makes cleaning more difficult.

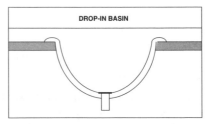

Drop-in basins can be used with any countertop but the lip is not user-friendly.

This lip on top of the counter may be even more objection-
able in the vanity than in the kitchen where there is always
a sponge or cloth around to pick up spills. In the vanity you
probably have to go searching for something to clean up hair
or beard trimmings that fall
on the counter. How nice to
be able to just brush them into
the basin. Again, the voice of
experience.

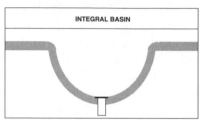

Integral Basins—When the
basins are integral with a cultured-marble counter-top, you
save money, both for materials
and installation. The disadvantage is that if you ever need to
replace the basin it's not
straightforward—it usually involves the replacement of the
entire countertop.

*With cultured marble, basins can
be molded with the countertop.*

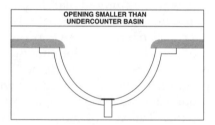

Undercounter Basins—In
the usual case the basins are a
little larger than the hole in
the counter so that the edge of
the counter sticks out over the
basin.

*Useful with any countertop except
laminate.*

Having had a cultured-marble counter with a separate undercounter, cast-iron basin, I would recommend it as a
way to get the advantages of the cultured marble counter without
worrying about what the basin may look like after several years.

A more sophisticated version is to have the hole in the countertop contoured to the exact size of the basin underneath (something
not possible with tile). This gives a very clean, smooth, and unified
appearance to the counter and basins. This appearance is enhanced
even more when the colors used in the counter are matched to the
basin underneath.

How basins are mounted under the counter depends on the type
of countertop material.

• Laminate counters, as noted earlier, are not suitable for undercounter basins.

A good-looking basin under a tile counter.

• Tile presents a special problem. With the small almost circular basins used in vanities, finishing pieces of tile must be made so short that they can give a busy unattractive look to the basin. With a relatively large basin and a very skilled artisan, it *is* possible to get a handsome undercounter basin in a tile counter.

Stone or cultured-marble counters with matching basins can be very attractive.

Beautiful marble and basin combo.

I've seen both. A poorly done basin in a tile countertop is ugly. A well done one will do justice to the finest bath.

- Stone counters, particularly marble with a suitably colored basin, is another attractive vanity. Because the color in stone is not just on the surface as it is with tile, it becomes possible to contour the stone countertop to match the shape of the basin resulting in a luxurious combination.

- Solid-surfacing veneers require basins that are cemented directly to the veneer for a seamless interface. Cast iron sinks may be too heavy for this. Or maybe you can use the technique described in Chapter 13. It'll depend on how comfortable your countertop fabricator may be with it.

- Cultured marble countertops are made with holes for the basins molded in them. The edges of the countertop around the holes are finished (gel-coated) and mounting tabs are molded into the underside of the countertop. These are used by the plumber when the counter is installed.

Banjo Counters

In the guest or hall bath, the countertop is sometimes extended out over the top of the adjacent toilet tank. While usually done with cultured marble, these banjo-shaped counters are sometimes used with other materials. It looks nice and adds a little counter space. But, if it is not to be an eventual maintenance headache, it takes a very low toilet tank. With standard tanks the counter makes it impossible to get inside them. Usually this won't be a problem for a few years until a valve needs replacing. Then it will be necessary to move the whole toilet to be able to get inside the tank. This turns what should be a small chore for the home owner into a major job where a plumber will be needed to reseat the toilet. *Caveat Emptor.*

Nice counter, but it blocks access to the toilet tank.

Pedestal Wash Basins

Half-baths with pedestal basins have become very chic in many parts of the country. But be sure to have a place where paper guest towels can be laid out because many people object to having to use a communal cloth towel, particularly in someone else's bathroom. The top of the toilet tank can be used in a pinch, but a small shelf close to the basin is better.

Bathrooms

Bathrooms with Two Doors

Hall or guest baths are sometimes made with entry doors from the hall and/or one or two bedrooms. The bath may be divided into a room with vanity and two basins and a separate room for shower and toilet allowing two family members to use the bath at the same time.

When the dual basins, toilet, and shower and/or tub are all in the same room it's not clear what the advantage is in having two basins because one person usually ties up the whole bathroom. Check this arrangement carefully when you run across it to see if it fits your needs.

When checking out bathrooms, the obvious things to look for are tubs, showers, and the number and locations of wash basins. But there are other design considerations.

Auxiliary Hot Water

Small hot water heaters are available that mount under a vanity counter to provide hot water immediately. If you want one for a shower or tub, the heater will need to be bigger than one that furnishes water to basins only. As with other hot water heaters, you'll need gas or electricity brought to the heater's location. The feasibility of doing this and the cost will depend on how big a heater you'll need and on the impact it may have on the rest of the house design.

Of course, recirculating hot water is another option. (*See* Chapter 6.)

Picture Windows in Bathrooms

A popular arrangement in master bathrooms is to put a big picture window right next to the bathtub. Or the tub may be in a corner so that there are windows on two sides. (I even saw a large clear glass window on one side of a shower stall—in California, of course!) These windows give a feeling of openness and often provide a great view. The question is, what kind of window treatment is appropriate? Sometimes it is possible to reach the controls for the blinds without taking off your shoes and climbing through the tub but more often than not it is simply downright user-hostile.

Here are some options:

1. Put permanent coverings over the windows to ensure privacy but you'll lose the openness that was so attractive at first.

2. Install mini-blinds that you adjust by climbing into the tub. If you find the right adjustment that allows you to let in some light while preventing people from seeing in even at night, then don't readjust the blinds.

3. Get blinds with motorized controls.

4. Use pleated cloth shades that close from the bottom up. Close them up far enough for privacy, but not so far that you block out all natural light. But you'll lose your view.

5. In one expensive, single-story house I was in the builder handled this problem by building a six-foot fence a few feet out from the window. Again, no view.

Other ways to let in outside light include using smaller windows placed high up on the wall or replacing windows with glass blocks, stained glass windows, or frosted window panes. You may want to consider acrylic blocks, which are less expensive than glass blocks, for both the material and installation. These are available in clear and several colors.

Glass block is good for light and privacy.

You can have light and a feeling of openness while maintaining privacy by having windows high up on the wall where they are met at the ceiling by skylights above the tub. If this appeals to you, be careful of the potential summer sun problem, particularly if the arrangement faces south or west.

Whirlpool Pump Access

Whirlpool tubs can be expensive, and you probably won't re-coup your costs when you sell the house. It's a good idea to talk to a few people who have them and get their reactions. Some people swear by them but many more say they're never used.

The CABO One and Two Family Building Code (1995 edition) says, "**3222.1 Access panel.** A door or panel of sufficient size shall be installed to provide access to the pump for repair and/or replacement."

Yet I've seen houses on both coasts where there is **no** access door or panel! And I've had a contractor's supply house tell me there is no requirement for it!!!

About as bad as no door or panel is to have one that's too small, in an inaccessible place, or located so far from the pump that it's useless.

The poor wording in the code (it doesn't say where the panel must be, just that it must be of sufficient size), the builder's inattention, and the lack of enforcement leave it very much up to the buyer to insist on something that's useful.

When the door or panel is put in place, there's no reason it has to be an eyesore, but it often is. The locations builders use for the access door are limited only by their imaginations. Both form and function are involved but there is absolutely no reason you can't have both.

Options for the location of the access door include:

1. A panel on the side of the tub. An aesthetic consideration. If done nicely this can be unobtrusive but not unless it looks as if the panel is an integral part of the tub skirt. It is also not acceptable if the pump is so far behind the access door that it cannot be reached.

A nicely-done pump access door.

This tile skirt would look better without the door in it.

2. In a room that backs up to the tub. Closets are the usual location for this option (good) but laundry rooms are also used (OK).

3. An adjacent bedroom where it can be an eyesore. Builders sometimes put the grill of a cold air return register over the access door to give it the appearance of being a part of the heating system. It's ugly. There are better ways to handle the problem.

A pump access panel in an adjacent closet is a good choice.

A less than attractive pump-access door.

If access to the pump is required, the toilet will have to be moved.

This inaccessible hole passed inspection!

4. In the room with the toilet. But be careful because it may be so crowded that it would become necessary to remove the toilet to work on the pump—a functional consideration.

5. In the inside of the end of the vanity cupboard that abuts the tub. Maybe it meets code but I can't imagine a plumber being able to get down under the basin plumbing and reaching through that little hole to do anything. Aethestically great—it doesn't show—but functionally very questionable.

6. An outside wall. This may be done even when it's on the second floor.

Outside is a good place both functionally and aesthetically.

Obviously good weatherproofing is required. It is, however, better than having an eyesore or an inaccessible arrangement in the bathroom.

Towel Rods

Many modern-day bathrooms are open and commodious. There are mirrors, doors, a tub with its surround, and impressive amounts of glass. But in the process of putting all of this luxury into the bath, designers and builders will leave no place for towel rods, again forgetting that "form follows function." Look carefully at the plans for the bathroom to be sure that there are places for towel rods, hopefully next to the shower, the tub, *and* the wash basins.

To me, using the tub surround as the place to lay towels when bathing is not satisfactory because they simply won't dry that way. It can be lived with. A little planning can avoid it.

Medicine Cabinets

Medicine cabinets are not only utilitarian, they can also improve the appearance of a bathroom—but they are simply not used in some regions of the country. If you want one in your house and you live where they are not common, you may have to use extra care to be sure that your wants get translated into reality.

Medicine cabinets are usually mounted in walls between studs. There should be instructions for the framers to leave room between studs to mount them. And be sure there is nothing else between the studs, such as pipes or wiring. Don't plan on using an outside wall for the medicine cabinet, because the insulation in the wall will prevent it.

An alternative to a medicine cabinet is to put a cabinet on the wall above the toilet. This option has a couple of disadvantages:

1) The toilet is in the way when you want to get into the cabinet and

2) If you accidentally drop something or brush it off of a shelf, it's likely to land in the water.

Can You Sidle?

Many master suites have the toilet in its own small room or in a room with the shower. When the room is too small, you have to sidle in beside the toilet bowl to have enough room to close the door. When the same arrangement is shown in floor plans, sometimes the toilet bowl is not to scale so the problem isn't obvious. One option is to leave the door off altogether. Another is to use a pocket door if appropriate wall space is available (no wires and no plumbing).

Usually it's not hard to avoid the sidle toilet when you think of it at the planning stage. It only takes a few more inches between the swinging door and the front of or beside the toilet bowl to eliminate the problem.

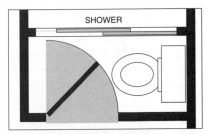

Sidle anyone? The problem isn't obvious when the door is missing in a model or a floor plan.

Exhaust Fans

Baths and showers should have exhaust fans even if not required by code. Fans can help reduce the amount of moisture which, in turn, helps control mildew.

Most importantly, get fans that are large enough. Codes allow ventilation in bath and laundry rooms to be accomplished either using a window or an exhaust fan. Most of today's homes include the fan. This is a good idea, even if there is a window, because in the winter you'd probably rather not open it.

Unfortunately, most bathroom and laundry exhaust fans pay lip service to the code but don't do much for ventilation. A fan with a capacity of 60 cubic feet per minute (CFM) will usually meet code requirements. For the subcontractor the cost of such a fan is around

$20. In a typical installation, a 4-inch flexible aluminum duct is tied to the fan to take the exhausted air outside the house as required by the building codes. At the lower end of the duct is a flap that keeps cold air from coming down the duct when the fan isn't operating. The fan must push air hard enough to open the flap then get the air up ten, twenty, or more feet of twisting, very rough duct. The $20 fan simply can't do this. And you get a useless exhaust fan.

A part of the reason that nobody seems to care whether a fan does its job or not is that there's a real question of just how useful they are in the first place. If you want to get odors out, then you'd better get one that'll deliver the 60 CFM (or more) when it's actually installed, not when it's sitting in a test lab someplace. If it's moisture you're trying to get rid of, you may ask yourself just how important that is in today's houses. If you want to give an impression that something is really happening then go along with the improperly installed $20 fan—they make enough noise to make everyone think they must be doing a great job!

Another place where fans are useless is in a modern open bath/ shower that's an extension of the bedroom. Here the area to vent is much larger. Fans may meet the letter of the code. They can't possibly do anything more.

On top of all that, note that houses are being made tighter and tighter in an effort to reduce energy use. This means that the fan has to fight not only the exhaust ducting, it's also got to fight to get air into the house to blow out! Should the codes be changed? You bet! Then maybe we'd get our money's worth in at least this one small part of the house. And it's not just the $20—the ducting and the installation cost the builder and you several times that much for each and every bathroom and laundry. And all wasted!

I've developed a method of quickly and unobtrusively testing these exhaust fans where I tie a piece of thin thread on the end of a stick and hold it up to the fan. In only two different houses out of perhaps 100 did I found fans that moved the thread. But in most bathrooms and laundries it never wiggles; the fan moves no air at all, none

Building officials know this but the industry has been doing it this way for so long that no one has the guts to try to change it. And homeowners, nationwide, are getting ripped off by most of the industry including code officials as well as builders.

What can you do about it? Make a point with your builder that you want a fan that moves air. It can be done but it'll be up to you to insist. (The trick is to use short smooth ducting.)

Heat

A heater in front of a shower or tub will make it more comfortable when towelling off. An overhead infrared lamp, a wall heater, or an electric baseboard heater all do a good job.

You should tell the builder the kinds of plain lights, lights with exhaust fans, heat lamps, and heaters you want. Fans, plain lights, heat lamps, and electric heaters should all have different switches, since not all of them are needed every time someone uses the room. And, if it matters to you, be specific about the exact locations of the elements and the wall switches that control them. The electrician will do the best he can but he can't read your mind.

Tubs

When you have a say, select a tub by manufacturer and model number—or let the builder do it for you and hope you end up with something you like in terms of style, material, and price. Your local plumbing supply store *should* be able to explain the pros and cons of the various available tubs, but not all of them can or will. It is probably prudent to visit more than one supply store and ask your questions. That way you stand a better chance of getting competent advice and of making the right decisions.

Some cultured-marble manufacturers make some really exotic appearing tubs that may appeal to you. These units are very heavy and the cost of shipping them is exorbitant but, if they are made locally, you may want to consider them. (If you buy one, be sure it's certified by the national cultured marble organization as explained above.)

Your choices for tub surrounds (the tub deck and the skirt in front) are ceramic tile, cultured marble, stone (marble), and, perhaps, solid-surfacing veneer. Don't consider plastic laminates, they make unattractive surrounds. Note that the material used for the tub deck may be different from the skirt but it's common to use the same material for the deck and vanity countertop. And, if you're using a whirlpool tub, don't forget the pump access discussed earlier.

Showers

While you're at the plumbing supply store, look at shower enclosures. Molded fiberglass units—both those with a gel coat surface and those made with acrylic—are popular. Cultured marble is another option. Since it has no grout, cultured marble does not support mildew growth, and its smooth surface is easy to keep clean. If you live in an area where cultured marble is not commonly seen in new houses, you may be surprised at how handsome it can look.

Ceramic tile is another popular shower enclosure. While mildew is not as serious a problem as it used to be, shower stalls need to be watched in the damper climates. Be sure that tile is mounted on a stiff waterproof board.

All shower stalls should have a hand hold in them. You need something to hold onto while you're standing on one foot washing the bottom of the other one. Don't forget to insist on a slip-proof bottom in the shower!

Most men wash their hair in the shower. The shower head should be high enough that they don't have to duck to get under it. This means putting the shower head in the wall above most manufactured shower stalls, i. e., well over six feet above the shower floor.

Chapter 21

The Living Areas

We have gone through the kitchen, the bedrooms, and the baths. We now look at the rest of the house: the entry, hallways, stairs, and dining room.

The Entry

The entry may be a separate room or it may be a part of a great room encompassing the living room, dining room, and/or family room. Stairs, if there are any, usually start from the entry. Although a large entry or foyer can add to the feeling of luxury in a house, it is an area that is not used that much. You may decide that you want one that is fairly small, saving the floor space for something more useful.

Occasionally a designer will forget to include a guest closet. That's usually not a good idea. When it *is* included, be sure it's reasonably close to the entry door.

The guest closet should not be blocked by an open entry door. It's just when people are coming or going that you need access to the closet.

Having a front door that opens across a staircase is almost as bad as one that blocks the guest closet. In general, you shouldn't have to close one door to open another or to use the stairs.

Like the chandelier over the stairs, be careful when there is a chandelier in the entry. Make sure there's a place to set a ladder to change light bulbs when needed. Entry lighting fixtures should be centered on either the entryway itself or the front door.

Although they are not as popular today as in years past, the split entry home is still a useful concept, particularly for a hillside lot. People considering these homes need to take a careful look—before the house is built—at what visitors will see when they come in the front door.

The arrangement shown here was seen in a recently completed home. The view into the hall bath is superb—what a great first impression *that* makes! Sure you could keep the bathroom door closed, but most of us don't—in fact we leave it open to let everyone know when it's not in use.

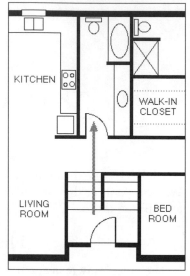

Now this may not bother you but it is a real turnoff to many people who feel that the view from the entry, or from any place in the living area of the home for that matter, should not include a bathroom. When you are considering floor plans, it pays, particularly at resale time, to avoid arrangements that may be aesthetically unappealing.

Split-levels require careful planning if you care about the view from the entry.

Floor

Even without separate walls, the entry is usually an identifiable area. This is accomplished using a different floor covering than adjacent rooms or, as is commonly done in southwestern architecture, by making the entry raised, requiring a step down to the rest of the house. Floor coverings need to stand up to water and mud, so carpet is rarely used. The most common entry materials include hardwoods, linoleum, ceramic tile, and marble and other stone. For safety reasons, be sure the material isn't slippery when wet. Surprisingly, there are floor tiles that are like ice when your shoes are wet and, unhappily, many decorating and tile centers are more interested in selling them than in warning you of potential danger.

Hallways and Stairways

These two areas are among those that are nonliving space. They are called "ways" for good reason. They neither hold furniture nor are they areas where we spend any time. In more extravagant house designs they can be eye-catchers or even dramatic, like a wide curving staircase or a bridge between two upstairs living areas.

Hallways

Look carefully at model homes to get a feel for how wide your hallways should be. A few inches can make a big difference in whether a hall feels comfortable or cramped. Hallways less than about 44 inches wide feel cheap and constraining to many people.

The hall in the drawing is in the lower level of a house that is dug into a sloping lot. The upper story is significantly larger than the downstairs. The bedrooms are on the open side of the lower story and the hall is on the uphill side. Even in daylight the hall tends to be dark. The particular house we saw did not have enough lighting nor was the ceiling high enough to get away from the tunnel effect.

People react to aesthetic things differently and what may be okay for you may not be for someone else. Things like an uninviting long hallway will make some people turn away from the house without realizing why.

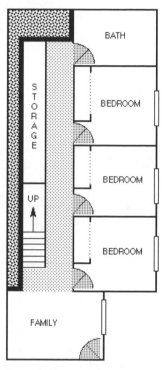

Long halls can give a sense of foreboding.

Lighting in Hallways

Most of the time, light fixtures in halls are placed where nothing in particular needs lighting. There should be hall lights where they're needed—in front of closets and the thermostat.

Stairway Design

Luxurious stairs are easy to climb but take a lot of floor space while steep stairs take less space but are difficult to use. In some designs, stairs are too steep for comfortable use. Check out one to see how you react to it.

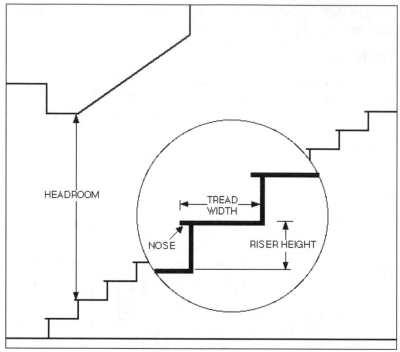

Stair design is important for code compliance, aesthetics, and ease of use.

A good angle for the stairs is about 36 degrees with a riser height of 7¼ inches. Each step width should be close to 11½ inches including a 1½-inch nose.

Building codes usually require a minimum stair width of 36 inches. You will probably want more. Similarly, codes require a minimum head room of 6 ft.-8 in. Again you should consider more, say 7 feet, to give a more open feeling.

Stairway Lighting

Have you seen chandeliers that require scaffolding to change a light bulb because there's no place to set a ladder? *Caveat Emptor!* Look for them when you're visiting models and looking at floor plans. Consider these alternatives:

• A chandelier with light bulbs pointing down that can be changed by a special tool on a pole.

• Sconces along the staircase and/or on the landing to light the stairs.

- Using several unobtrusive nightlight-type fixtures that are mounted in the wall a couple of feet above the treads in the stairwell. These units may be small (the size of a standard outlet cover plate) or they may be two or three times larger. They're installed flush with the staircase wall and have louvers in them to direct the light downward.

- Have recessed can lights installed in the ceiling above the stairwell. This may be the least attractive solution because when you look up the stairs you look straight into the lights.

Chandeliers over stairs can be maintenance headaches.

The Dining Room

In small starter houses the formal dining room is often omitted. The idea is that a young family has better uses for the space than for a room that seldom gets used and requires a significant outlay for furniture. If you decide to do without a dining room, be sure the kitchen eating area is large enough to accommodate your family and guests.

Today's formal dining room isn't always a separate room. Often, it's an area in a great room set aside for the dining room furniture. Ways to distinguish it include using archways, partial walls, columns, changing the ceiling height or pitch, flow of a wall, the pattern in a heavily textured ceiling, or a different wall paper.

Designers and builders hesitate to commit any more space to the formal dining room than they can get away with. Home buyers recognize that dining rooms aren't used all that often, so they're torn between saving money and making the room big enough to accommodate a full-size set of dining room furniture. Unfortunately, the answer to this problem in many of today's designs is to include a formal dining room in the floor plan and then make it so small that it won't take a set of furniture.

Room Dimensions and Furniture Placement

When we moved into our tract house in southern Califor nia, we had a separate formal dining room that barely held our dining room furniture. When we decided to replace our old furniture with a nicer set, there was no longer enough room. Whenever we entertained it was difficult to pull out the chairs so people could sit down and almost impossible to get behind chairs with people sitting in them.

At least we could extend the table out into the family room to seat everyone when we hosted the family Thanksgiving dinner.

Our present dining room doesn't use much more floor space but it's open on two sides. What a difference. We have room behind the chairs and enough space to expand the table for large gatherings.

When deciding on the size of the dining room you need, remember to have room for a full-sized dining set in the future even if your present dinette set would fit now. And then, when you sell, you won't turn away buyers who want a room big enough for their dining room furniture.

Wall space in a dining room is a precious commodity. There must be room for a hutch or china cabinet. Some dining room sets include a server and there must also be room for that or a sideboard along a wall.

Sometimes builders will build in the china cabinet so that you don't have to allow for it in the dining room layout. Take a careful

This dining room is too small. You should be able to comfortably get behind the chairs when people are sitting in them.

look at such designs because it's easier to pay lip service to "built-in china cabinet" than it is to provide one that's large enough. The drawing shows such a situation. The corner cabinets will hold only a fraction of the china and glassware that a full-sized china hutch will. *Caveat emptor.*

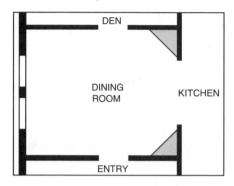

These corner units are meant to take the place of a china cabinet. They don't.

The table is usually put near the center of the room and the dining room arrangement should be able to accommodate extension of the table for large parties. This is often into an adjacent room or area.

A dining room table is typically about 42×72 inches. Another 36 inches of clear space on each side of the table is needed so that there is room to walk behind the chairs with people seated at the table.

A typical china cabinet is 22 to 24 inches deep so that the minimum dining room width becomes 11 feet 4 inches. The popular 10 foot-wide dining room just isn't enough.

The length of the room without additional leaves in the table and with someone seated at each end of the table needs to be 12 feet. If there is no way to extend the table into an adjacent room or area, then the dining room should be 32 inches longer.

Any doors that extend into the area occupied by the table and chairs also need to be considered.

This is a lot of room for a small house. Clever combining of the dining room with an adjacent room can cut this down significantly without compromising the amount of room needed for the table and chairs. With dining rooms that have three or four walls, it is not possible to take advantage of the presence of the adjacent rooms.

Watch out for sunken living rooms. A dining room table cannot be extended in that direction. Unfortunately, it is easier to remember this after you've already moved into your new home.

The Chandelier

One of the more disturbing things seen in many of today's houses is the inattention paid to how the furniture will look in the dining room. In particular, the chandelier should be centered over the table—to do otherwise is visually jarring. Unfortunately that is something which many times cannot be fixed after you move in. Planning is critical if you do not want to live with this aesthetic irritation.

Some contemporary home designs have forgone the traditional dining room chandelier in favor of recessed ceiling lights thus avoiding the problems with chandelier location discussed here.

The Chandelier and Furniture—In all but extra wide dining rooms, when there is a china hutch on one wall, the table must be centered between the hutch and the other side of the dining room. Then, to have the chandelier centered above the table, it, too, must be offset relative to the sides of the room.

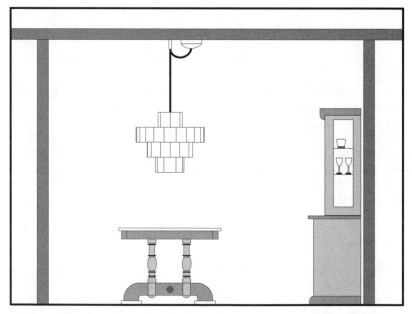

A swagged chandelier is needed when the electrician installs the mounting box in the center of the room.

If the electrician has not been thinking ahead or you don't make a point of it, you'll end up swagging the chandelier as shown in the illustration. Swagged fixtures are almost always the result of not thinking ahead. Don't let it happen to you. Tell the builder what you want before the electrician does his thing.

Ceiling Treatment—Be wary of special ceiling treatments such as coved, trayed, coffered, Pullman, a pattern in the ceiling texturing, or any other arrangement that is symmetrical around the center of the room. These all dictate that the chandelier be centered in the pattern, i.e., at the room center. If the room isn't large enough to allow the table to also be centered, you'll have the problem discussed above, only this time you can't swag the chandelier to fix it. These ceilings should only be used in wider dining rooms.

High Ceilings—Many homes in the Southwest have a similar problem arising from the high ceilings used to get the open airy feeling in the architectural design. In this case the chain holding the chandelier is quite long and, visually, it becomes an important part of the dining room. Because it is so obvious, it needs to be symmetrical with the tall room walls. Putting it off center simply doesn't look right.

So, again, we have a chandelier which is centered between the walls—a fixture that should only be used with a dining room table that is also centered which, in turn, requires a wider room.

Minimum Room Sizes—If there is no chandelier or if the chandelier can be hung off-center, room size requirements are:

• With the hutch parallel with the table: 11 ft.-6 in.×14 ft.-8 in.
• With the hutch at the end of the table: 9 ft.-6 in.×16 ft.-8 in.

If there is a special ceiling pattern or high parallel walls in the dining room, be sure your room is big enough to let you put a china hutch along one wall and still center the table under the chandelier.

Required room sizes then become:

• With the hutch parallel with the table: 13 ft.–6 in.×14 ft.-8 in.
• With the hutch at the end of the table: 11 ft.-6 in.×16 ft.-8 in.

Windows in Dining Rooms

Many floor plans have a window in the dining room. Be careful if it is close to being centered on the dining room table.

I n one furnished model home the dining room table was parallel to the wall with the hutch. The chandelier was hung in the center of the room and a window was centered there too. The end of the table was close to the window. However, because of the china hutch, the table could not be placed in the center of the room so that it was about a foot off from the chandelier and also about a foot off from the center of the window. The asymmetry of the table and the window was probably more visually irritating than the table and the chandelier.

In any case, be careful of arrangements like these. If there is a window in the dining room, take care that it is not located where it looks as if it were meant to be centered on the table but never quite made it.

The Family Room

If you are going to have a gas fireplace, the combination of the location and type of fireplace becomes important as discussed in Chapter 8. Besides that, you need to be careful where the fireplace is relative to the TV. The fireplace and the TV should be close to each other so that the same furniture arrangement is used to enjoy both. But you don't want the TV where a window can cause a glare on the screen during the day.

Ceiling fans are common in family rooms and the suggestions in Chapter 5 about running 3-wire power conductors apply here also.

Chapter 22

Laundry and Garage

The two utility areas, the laundry and the garage, are the subjects of this chapter. They have a potentially common thread, the utility tub—if there is one—that may be located in either place.

The Utility Tub

Many builders and designers do not include a laundry or utility tub in their houses and most home users want one. If you have a choice, you'll need to decide where to put it.

For pre-soaking especially dirty clothes, the tub is more convenient in the laundry. For cleaning tools and for cleanup after yard work and after play, the tub is better off in the garage. In the garage the faucet on the tub can also be used as a hose connection.

If your house plans permit a special heated "mud" room for the utility tub, that's even better.

The Laundry Room

Energy

If there is an option, you will have to decide early whether to plan for a gas or an electric dryer. This is often a regional consideration; gas is the norm in some areas and electricity in others. Even when you start with electricity, having the laundry also plumbed for gas will make it possible to change over to the less expensive energy in the future. It may also make the house more attractive to buyers when the time comes.

Location

The laundry can be a simple niche off a hallway, it can be a separate room including ironing board and a broom closet, or it can be missing entirely with the washer and dryer in the garage. Whichever arrangement you choose, here are some things that should be kept in mind:

• Washers and dryers are noisy. Consider the location carefully. Think about noise insulation. Don't use louvered doors.

• Dryer doors open downward or they are hinged on the side. Some allow the hinges to be moved to either side, others do not. If you have one with side hinges that cannot be moved, locate the dryer so the dryer door isn't in the way when moving clothes from the washer.

• Put the washer and dryer right next to each other. Separating them will mean extra steps.

• If you have a tub in the laundry room, put it next to the washer. If you have a clothes-folding area, put it next to the dryer.

Common,
thoughtless
arrangement.

• Figure out where the dryer vent is going. As discussed below, it can be a shock to find it sticking out the front of the house.

• Be sure there is a light over the washer that illuminates the interior of the washing machine even when you are leaning over it. This is a common shortcoming, one you'll appreciate when you try to find a dark sock inside an unlit washer.

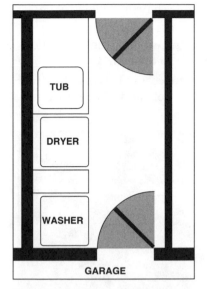
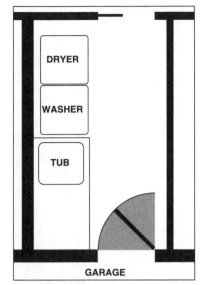

Common, but poor, design.   *Good design.*

When the laundry is the hallway between the garage and the rest of the house, there are several problems to watch for. In the drawing on the left:

• Anyone working in front of the washer will be hit in the side by the door knob if someone comes through the door. **This hazardous situation should be avoided**.

• The washer isn't right next to the dryer.

• The tub isn't next to the washer.

By using the arrangement in the right figure, these problems are avoided. Note the use of a pocket door into the rest of the house to avoid a swinging door that could cause problems for someone using the dryer. You could eliminate this door entirely if it is into a hallway or someplace else where noise from the appliances will not be a concern.

Laundries with Corners

In general, avoid laundry rooms with the washer and dryer across one end and a counter along the side. These will leave an area under the counter that is inaccessible when the washer and dryer are in place. In the photo, where the machines haven't been installed, it isn't so evident that:

- The tub will be well separated from the washer (the connections for the dryer, not the washer, are next to the counter) and

- The space under the counter will be inaccessible when the dryer is in place. Watch for this in both models and plans; it isn't always obvious.

Corners in laundry rooms are inaccessible. (And don't let the clothes basket fool you.)

Other Amenities

A well-designed laundry will have:

- Cupboards over the washer for holding laundry materials (soap, bleach, softeners, etc.),

- A short rod over the dryer for hanging clothes, and, if there's room,

- A counter for folding clothes, and

- An ironing board.

If you have a two-story house, here are arguments about the laundry room location:

1. If the laundry is on the first floor, you should have a laundry chute to get the clothes to the machines. Then you only have to carry things back upstairs.

2. If it's on the second floor, it'll take fewer steps to collect the laundry and to put it away but it'll take more trips up and down the stairs while the washing is being done to take care of adding fabric softener if you want to, transfering the laundry from the washer to the dryer, and checking if things are dry. And there is

always the possibility that the washer or its hoses will spring a leak causing much more damage than if it's on the first floor.

On this subject, I'll always remember a building designer's own new home that was on a tour and was praised as a model of excellence. The bedrooms were on the second floor and the laundry in the basement. There was no laundry chute! And this was not a cut-every-corner house, it had many upper-end design features. It's bad enough to have to carry laundered clothes up the stairs, but to carry them down, too? It was truly, as a woman real estate agent commented, a case of "it was built by men."

As a Mud Room

If you plan the laundry room is to double as a mud room, be sure of two things:

• It should not be necessary to go through a room of the house to get from outside to the mud room, and

• There should be a place to put dirty, wet clothes and to clean up before going on into the house. This strongly suggests a utility tub in the laundry and room for a hamper.

The Dryer Vent

The dryer vent may, at first, seem like a trivial problem that any mechanical contractor can take care of. *It isn't!* Unless having the dryer vent on the front of the house doesn't bother you, then pay close attention to where the laundry is located. If not accounted for, by the time the HVAC contractor gets into the act the only place for the dryer vent may, indeed, be the front of the house. And there are other aesthetic and functional aspects you should be aware of.

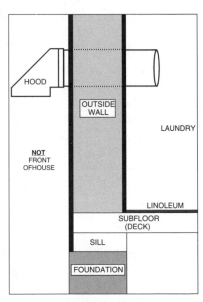

The optimum dryer vent is straight out the wall.

Dryer vents involve important design considerations which are universally ignored—until maybe the building inspector catches up with it. He, however, is not concerned about how it looks, that's up to you.

There are two design facets that need to be accounted for:

• Don't put the vent outlet on the front of the house unless you want the surrounding shrubbery strewn with lint. The same consideration says not to vent it through the roof.

• Make the vent pipe short and straight or it can seriously degrade the performance of the dryer or even cause the dryer motor to overheat and start a fire. And, if the air flow is reduced, lint can build up in the duct—another place for a fire to start. (Because of this there are new building codes that mandate the maximum length of the duct.)

It is this potential of fire that has made code agencies take a harder look at dryer vents. The dryer, in effect, is a big blower that heats air and blows it through the clothes. The exhaust air is laden with moisture and with the fine lint that gets past the built-in lint screen. Anything that impedes the air flow degrades the performance of the dryer. The duct should terminate outside so the moisture and lint don't mess up the house.

There should be a hood cap with a flap or some other way to prevent rodents and birds from getting into the duct. If a hood cap points downward it should be at least 12 inches above the ground for unimpeded flow.

Section 1801 of the 1995 CABO model code says, "The maximum length of a 4-inch diameter exhaust vent shall not exceed 25 feet from the dryer location to

The white square in the lower right corner is the dryer vent —on the front of the house!

wall or roof termination...A reduction in maximum length of 2.5 feet for each 45-degree bend and 5 feet for each 90-degree bend shall apply. Installations when this length is exceeded shall be installed in accordance with the manufacturer's installation instructions."

When building inspectors take these words literally, they'll insist on using the CABO values unless the actual dryer is in place at the time of inspection (in which case they'll consider the manufacturer's instructions). If care is not taken and the dryer is located toward the front of the house, going out the front may be the only option.

W hen a building designer was told that he must now take this into account his reaction was, "Now you're telling me I must design the house around the laundry room? No way!"

Yet that is exactly what must be done. It's another constraint that should have been in place for years. Be sure you pick a floor plan where the dryer doesn't have to be vented to the front .

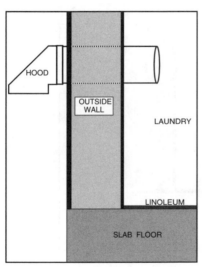

With a slab floor, you must put the dryer on an outside wall.

Whether in the wall or in the floor, insist that there is room around the end of the vent pipe for the connection to the dryer.

We look here at some ways dryers can be vented. Note that in *every* case, the best arrangement is to have the dryer on an outside wall (not the front) and run the duct straight out from the back of the dryer.

Slab Floor—When your house sits on a slab floor, the duct goes straight out through the wall or through the roof. To avoid the roof vent or running it through the garage, the dryer *must* be placed on an outside wall. Be sure this isn't the front of the house.

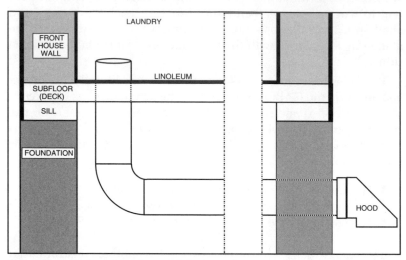

One elbow in duct. Total run:20 feet maximum.

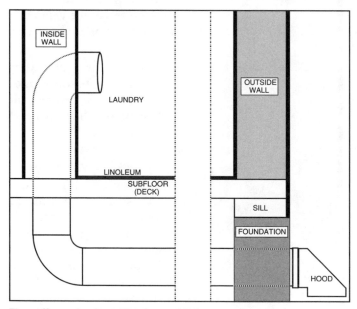

Two elbows in duct. Total run:15 feet maximum.

Outside Wall—When the laundry is on an outside wall—but not the front of the house—vent it through the wall. If it is on the front of the house, run it down through the floor and then to the outside under the floor. For this, the dryer will have to sit out farther from the wall and the linoleum must have a hole cut in it, making it harder to install and keep tidy. The CABO code allows the total run to be 20 feet long (one 90-degree elbow). Look for this situation and you may want to rearrange the laundry to avoid putting the dryer where it would have to vent through the front outside wall.

Inside Wall—There are two options:

- As described above, the duct can run straight down through the floor and then outside, with a total run length of 20 feet.

- The duct is run down through the wall (which must be 6 inches wide to allow room the for the duct) and to the outside. The CABO code allows a total of 15 feet for this run (two 90-degree elbows). It is imperative that this be accounted for when the plans are drawn so that the duct isn't too long—otherwise the vent might have to be on front of the house.

Laundry in Basement—The dryer should be on an outside wall where the vent duct can be run up the wall and outside.

Laundry in Garage—The dryer should be on an outside wall or the duct will have to be run across the garage to get outside.

A heating contractor recently showed me a set of blueprints made for a custom house by a competent, well-respected designer. The laundry is a small room that sticks out into the 3-car garage and its floor is a continuation of the garage floor, i.e., it is a slab. The only places to take the exhaust duct are up to the roof or into the garage. Neither is acceptable. [I didn't find out what happened. Suffice it to say that you can't be too careful—even the experts make boo-boos.]

The Garage

Garage Design

Finishing (wallboard and painting) the inside of the garage is something for which there is no standard practice. For energy code and fire code reasons, common walls between the house and the garage will always be insulated and wallboard installed—as will the garage ceiling when it is also the floor of an upstairs room.

Beyond that, you should specify just what you want done: what's to be insulated, and to what R-value, and what's to be covered with wallboard and how it's to be finished.

Garage doors (the big ones you drive your car through) may be metal or wood. The metal ones require less maintenance and last longer but are more expensive. Wooden doors that are only a year or two old but face the direction of the sun and of winter storms usually are in need of a coat of paint.

Whether and how many garage door openers are to be included by the builder is an item that you should specify, otherwise you may not get any because for this, too, there is no standard.

Unless the garage is just stuck onto the end of the house, the size of the garage in a given design is pretty well fixed. You should check if it's adequate for you. A difference of two feet in width and/or length can easily make the difference between being crowded and having room to put in a work bench and storage cupboards. As a starting point consider that less than 22×22 feet will be a crowded double garage.

If you want, your builder probably can change one double garage door to two singles or vice versa.

If your plans call for a window in the garage, or if you want one there, be sure that it is placed where it can't be used to provide easy access for burglars. Or, have bars put across it to block entry or make it of glass that cannot be broken for access. Such windows are usually on a side of the house that isn't readily visible, making them favorite candidates for someone looking for the safest way in.

Codes in some areas require that physical barriers be placed in front of the furnace and water heater to prevent a car from inadvertently running into them and starting a fire. Be sure any such barriers in your garage are firmly embedded in the concrete and are not just lip service to the code.

Garage Wiring

If you are planning an automatic sprinkler system for your yard, the usual place to put the controller is in the garage. If the garage is to be finished, the wires from the controller should be placed in the walls before the wallboard is installed.

Check the electrical schematic for outlets in the garage. Some houses are designed to have only the one that is required by the

National Electrical Code. You will want at least one on each wall. These can all be on the same 15-ampere circuit and must be GFCI protected. If you have a freezer or other appliance in the garage, you should put it on a separate circuit and not reduce the available capacity in the other outlets. It is a good idea to GFCI protect this outlet, too, even if it is not absolutely required by code. It is appropriate to specify exactly what you want and not leave it to the electrician and building inspector to do things right.

There should be power going to outlets for the garage door openers. Be sure to check, sometimes it's not included. It is convenient to have a switch in this line to shut off the openers when you are on vacation. (*See* Chapter 11.)

In an upscale tract in Montgomery county, Pennsylvania I saw what to me is a marketing outrage. Not only are there no garage door openers, there is no power wiring to where you would want to install them! It might be an oversight, but that seems unlikely from the size of the tract and the evident experience of the builder. The unwary will get seduced by the low price on the houses only find to their dismay that that low price vanishes when things like this—the wiring for the openers as well as the price of the openers themselves and their installation—are added on. Personally if I were house hunting and ran across something like this, I'd get out of there so fast that the smooth-talking agent would wonder what happened.

Garage Lighting

In some places it is customary to put one light socket in the garage for each car that may be parked there. In others, one socket is supposed to serve the whole garage. If your plan shows only one, then increase the number of sockets so that there is a modicum of light even in the dark of night. You may want more than one per car stall or you may want the electrician to install fluorescent shop lights for you or at least put in suitable sockets so you can install them yourself.

If you have ever come in the side garage door of your house at night and stumbled around to get to the light switch at the house door then you'll appreciate the suggestion that you have the garage lights wired with two switches in a 3-way operation, putting the second switch at the side door. (If it's too late for this, use a night light in the garage, it's a shin saver.)

Garage Sidedoor

Some designs show a garage sidedoor and others do not. There should be one unless the garage is dug back into a hillside so there's no place to put the door. The door may be metal or wood but, as with the big doors, metal will last longer and appearance is not critical. The door hinges and the dead bolt striker plate should be mounted with long screws for security reasons as discussed in Chapter 11.

A sidedoor with a window in it is a further security consideration. If you are thinking of such a window, Chapter 11 is worth rechecking.

Garage Floor

With areas of concrete as large as a garage floor there will be cracking when the floor settles. To control where this cracking occurs, the concrete subcontractor puts dividers in the floor before the concrete is poured. These plastic dividers (about one-half inch thick and four inches high) are set on edge just below the surface of the concrete. If a trowel is used to make a line in the floor above each divider, the inevitable cracking will occur at the bottom of the line. When this line is not put in the concrete, the cracking will be in a jagged unattractive line the length of the garage. Besides its appearance, this is hard to keep clean. Insist on proper finishing. And you'll need to be there when the concrete is being finished, otherwise it'll be too late.

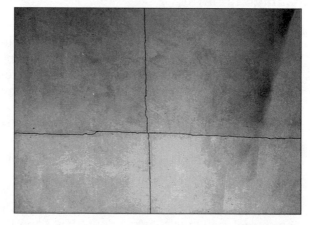

Heedless, sloppy work.

Chapter 23

Outside the House

Unlike the interior, functionality plays a role in only a few of the choices involving the outside of the house. Mostly it's a matter of cost versus looks. The options are reviewed in this chapter.

Entryways

There are two types of entryways you may want to avoid. One leaves the front door with no protection and the other puts the door at the end of a tunnel.

Protect the Entry

Here are three reasons for having a roofed porch over the main door:

- It provides a covered place for guests to stand when they ring the door bell and for you when you're fumbling with your keys.
- It keeps the sun and rain from quickly deteriorating the door itself.

A bare door is not an inviting entry.

• It gives the house more curb appeal. A bare door does not exude a feeling of "welcome."

The covered porch is not a part of colonial and federal styles and it is also absent in some contemporary and Southwestern architectures. It is, again, a question of aesthetics versus function. If maintaining the facade is imperative, consider a vestibule where visitors can be out of the weather while not inside the house proper.

The effects of weathering were seen dramatically in a tract in Riverside, California where, after only one winter, the finish on doors had broken down and the wood underneath was starting to come apart, particularly on the bottom third of the doors where they were exposed to both sun and rain. (Hopefully, the eventual owners got metal or vinyl front doors that stand up better to the weather.)

The Tunnel

Ranch-style houses with the garage facing the street often put the door at the end of a long entryway that runs between the garage and a house wall. With roof overhangs, this arrangement has all of the appeal of a dark tunnel.

The worst—an unprotected door at the end of a tunnel.

Exterior Materials

A first consideration in deciding what materials can be used on a house is the CC&Rs. Many of them will spell out what is possible and what is not. Even if they're decades old and out of date, they're still the law of the land where your lot is concerned. (Appendix A is a tutorial on CC&Rs.)

The exterior material you have on your house will depend on your personal taste, cost, the general availability in your area, its suitability for your climate, its consistency with the house's architecture, and its congruity with the neighborhood.

To a large extent the question, again, is a matter of aesthetics versus cost. Inevitably the most expensive is perceived as the most desirable, which may or may not be the case. But, when resale is an important consideration in your decisions, it becomes advisable to choose a material that will make the house attractive to the greatest number of potential buyers.

Your builder or local building supply house will be able to tell you the available options and their relative costs. It isn't a bad idea to have some idea about these even when you're looking at tract and spec houses because it gives you a better feel of the quality of the house.

You'll find the use of actual wood is infrequent. Builders prefer vinyl or a composite board or hardboard that's made to look like wooden boards, shake, or siding. This material virtually eliminates the problems of warping and splitting that can raise havoc with boards and shake. In general, it is also less expensive.

Stone and brick are not used extensively in earthquake-prone areas as the primary exterior material because of the cost to reinforce them.

When every effort is being made to keep costs down, one material may be used on the front of the house for appearance and a less expensive material for the sides and back. For example, many houses in northern California use a board-like hardboard material on the street side and stucco elsewhere while in the Northeast, facades may be stone, stucco, brick, or vinyl siding depending on the architectural style being portrayed. Vinyl is used extensively on sides and backs.

Stucco is used almost exclusively in the Southwest, with occasional wood, brick, or stone accents. In damper climates, some stucco is used but there is concern that it may develop cracks with time, letting water into the wall space that would never dry out. To overcome this problem, stucco-like synthetic materials are used.

One problem with stucco is that it stains easily. This was seen in a tract where the two-story houses had no rain gutters, very small eaves, and little landscaping. Water running off of the roofs splashed mud onto the stucco and permanently stained it. (The builder had gone bankrupt—imagine that!) It's important, particularly with stucco, to get gutters on the houses as soon as possible to prevent this from happening.

Roofs

Rain Gutters

It's a good idea to have gutters included with the house when it's built, regardless of where you live. They are not used in some parts of the Northeast because of the perception that snow clogs them up and can cause water to come into the house. As explained below, this can also be caused by inadequate overhangs on the house, gutters or not. If you feel more comfortable about it, leave the gutters off, it's certainly less expensive.

In parts of the Southwest, builders rarely include rain gutters as a part of the houses they build. In other areas no one even thinks about them because they're always included with the house. A set of gutters can run well over $1000, so don't dismiss them as trivial when you buy in areas where they're not ordinarily included.

Gutters may be plastic, aluminum, or heavily galvanized and painted steel. They are available in two sizes; a larger one that completely covers a 6-inch fascia and a smaller one that covers part of the fascia. Another case of cost versus appearance.

A money-saving strategem builders use in many places, when they do install gutters, is to just dump the water from the downspouts into the yard. Unless the soil is particularly porous, this will cause a muddy mess. It's not a big thing to have drain lines included if you have a downhill shot to the street. Having them is one less thing to worry about after you move in.

Roofing Materials

There are seven commonly used roofing materials:
- Concrete tile or shake
- Composite tile or shake
- Clay tile
- Cedar shake
- Woodruf, a hardboard material that simulates shake
- Composition roofing
- Asphalt shingles

Concrete Tile—Concrete tile or concrete shake comes in many shapes and colors. It has the longest life of any of the materials. It is not damaged by weather or age but it will break when walked on. (It can also be broken by golf balls if you live on a golf course.) It is, obviously, fireproof. It is the most expensive roof, not only because of the cost of the material itself but, being concrete, it's heavy and may need a sturdier roof structure to support it. Its weight and the ease with which the pieces can be broken make it more difficult to install, all adding to the cost.

Composite Tile or Shake—This material looks like concrete but includes fiberglass or other strengtheners to reduce its fragility. It is lighter than concrete, making it less expensive to install.

You'll usually get back some of the initial extra cost of nonflammable materials through lower insurance rates. This is something you'll need to check. In general, the ":carrots" for using concrete tiles or concrete shakes are fire resistance and near-zero maintenance costs.

Clay Tile—Clay tile is a low-fired brick-like material. It is usually made in the form of a half of a red tile pipe. It has been used for several hundred years in southwestern architecture but has given way to its concrete counterpart because of its fragility. Clay is even more subject to damage from being walked on or hit with golf balls but is less expensive than concrete.

Cedar Shake—Split cedar shake has been a popular roofing material for a long time. It's perceived as being the luxury material in many areas. Unfortunately, the quality of available shake has declined. Now it's the shortest-lived modern-day roofing material except for the cheapest asphalt shingles. It requires more maintenance than any of the others. It's very flammable ("tinder box" is a term that is often used) which results in higher fire insurance premiums. It's more expensive than other materials except concrete tile. Yet, in the Pacific Northwest it's still very popular. CC&Rs in many exclusive developments allow only cedar shake, concrete tile, or concrete shake roofs for appearance reasons. In these developments, cedar shake is seen much more often than concrete tile. Obviously, personal taste figures heavily in this matter. This is unfortunate because cedar shake is, by far, the worst buy in roofing materials available today.

Woodruf—A material with the trademark "Woodruf" is manufactured by the Masonite Corporation as an alternative to shake. It's a composite wood product with an appearance similar to shake, but more uniform. It's less expensive than shake and is often permitted where CC&Rs require shake or tile. The expected life of a roof of this material (15 years) is about the same as a cedar shake roof but not nearly as long as concrete tile or the better grades of composition roofing. It, too, will burn but not with the same vigor as cedar shake.

Composition Roofing—Composition roofing is an improved, fiberglass-reinforced version of the old asphalt roofing that has been around for many years. It's a flat material that's installed very quickly using a staple gun. It comes in a number of colors and styles to match the house. It's less expensive than most other roofing materials and is available in a wide range of prices and guarantees (up to 40 years). It doesn't burn as readily as shake and walking on it does not harm it as much as other materials.

Asphalt Shingles—Asphalt shingles are the least expensive material and are short-lived. They're seldom used on new homes.

Your builder can go over the costs of the various options with you. Whether you go with concrete tile for its durability, shake for its appearance, or composition roofing for its cost is a personal choice. But don't forget perceptions, because they can be very important when you try to sell your house.

In a tract near Sacramento, California it was noted that the first phase had cedar shake roofs, the second had composition roofing, and the newest phase used concrete shake. The builder explained that after the Oakland fire storm, nobody wanted anything to do with cedar shake and that composition roofing is simply not accepted, because it's viewed as a cheap material used mostly for replacing older asphalt roofs.

Unlike the rest of the country, the only places where composition roofing was seen on new houses in California were on the lowest-priced starter homes. Regional perceptions can be strong indeed, often flying in the face of good sense.

In some areas, the under sides of roof eave overhangs (soffits) are commonly left open and in others they're closed in. When vinyl siding is used, vinyl pieces are designed to close in the soffits. With other siding, soffits are closed in with plywood that is fitted under the overhanging roof beams. This improves the appearance since it gives a much smoother finish to someone looking up at the underside of the eaves. Closing in is much appreciated when you get around to repainting your house. It is a matter of local custom, cost, and your personal preference.

Roof Slope

The slope of the roof is measured in "pitch" using an expression such as 4/12 which means there is a 4 inches of vertical rise for each 12 inches measured along the horizontal.

In today's houses there is less variation in the slope of roofs than in the past. There are two reasons for this: some roofing materials aren't guaranteed if the roof isn't steep enough and very steep roofs are not popular with builders because of safety concerns for installers. These concerns translate to higher costs for building steep roofs because of the care needed to prevent the roofers from falling and of higher worker's insurance costs.

Overhang

Roof overhangs protect at least part of the house from the direct rays of the summer sun, helping to keep it cooler and provide some protection from rains. In snow climates, roof overhangs serve another important function: they help keep melting snow water from seeping into the house.

During cold weather, snow on a roof will be melted by the heat of the house under it. The water runs under the snow toward the eaves and gutter. An ice dam can form at the roof edge backing up the water and causing the roof to leak. If there is inadequate overhang this water goes into the house, otherwise it will just seep through the overhang. (Further insurance against this type of leak is to have waterproof "ice guard" installed under the shingles just above the gutters. This material extends about three feet up from the gutters, preventing leaks caused by the ice dam.)

Roof overhangs are sometimes omitted for reasons of appearance in a particular design and are sometimes made too small to save money. If your house is in an area where it snows, make sure there's a substantial roof overhang, particularly with roofs with lower pitches. (24 inches is suggested as a nominal value.) In the South and Southwest much larger overhangs are useful because of the shade they provide against the summer sun.

Valleys

When there is a change in direction of the roof, either a ridge or a valley will be formed—ridges on outside corners, valleys on inside ones. Inept installations at valleys can result in leaking roofs, not immediately, but several years later after the builder is long gone.

With composition roofs the easiest and least expensive way to make a valley is to simply bend the roofing material through the valley and to lay the material from the other side on top. The top layer of material is then cut in a straight line up the valley to give a finished appearance. If done properly, this is fine.

If not installed properly, rain can be blown up under the edge of the top layer of roofing material. Over time dust and particles from the roofing will also be blown under there and, sooner or later, water will accumulate and leak.

There are three ways to prevent this:

• Be sure the roof is laid properly. This is hard to assess unless you're a roofer. Many builders wouldn't know the difference either.

• When the roof is laid initially have roofers weave the two layers of material, alternating them so that there isn't a single long line of cut roofing for rain, etc., to enter.

• Use a piece of specially bent metal in the valley (called a "valley W") that extends well back under the roofing material on both sides. A ridge in the metal keeps water from being blown under either side.

It took the fifth winter before our roof developed a leak. The cost of correcting the builder's lack of supervision was considerably more than it would have been if it had been done right in the first place.

This valley leaked.

This one did not.

Nooks and Crannies

Where roofs from dormers meet the rest of the roof there's usually a small protected area that's attractive to birds looking for a place to set up housekeeping. If this is okay by you, fine. If you'd rather the birds took their dirty nests, noisy nestlings, and multitudinous parasitic pests to some tree or bird house, then these protected nooks and crannies in the roof should be blocked off, either with wire mesh or boards.

This can be done at any time. However, it's a good idea for the builder's crew to do this so that you don't have to shoo the birds away when they've already decided to set up nest making. (In some areas it's illegal to do anything once nesting activities have begun.) This is yet another small detail you can have the builder do or wait and take care of it after you've moved in.

Concrete Work

Sidewalks

Front sidewalks can be utilitarian and ugly or smooth and flowing with lots of curb appeal. If you don't speak up, you're apt to have a walk that doesn't do a thing for the appearance of the house.

Avoid a house with a steeply pitched walk if you live where there's ever snow or ice.

Steps are better than a steep walk. But be careful with steps. If there's a possibility of someone not seeing the step, better have it well lighted and marked with white paint to avoid accidents. If

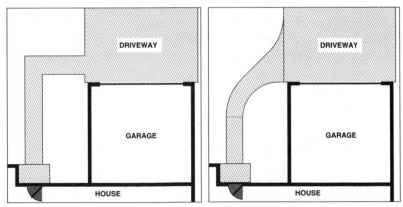

Different front walks (right-angles are easier for the contractor).

you are concerned with access by people with disabilities, you'll need a gently sloping walk without steps.

Conduits Under Concrete

When sidewalks, driveways, and RV pads are being poured, be sure you have enough conduits put under them. An irrigation system may eventually need to have a pipe go under a piece of concrete. It's a lot simpler (and cheaper) to put an oversized pipe (conduit) under the walk or driveway before it's poured. You can then put your irrigation pipes through the conduit whenever you need to. Alternately, the concrete subcontractor can lay a piece of ¾-inch PVC under the concrete for your irrigation system. This is simpler but lacks flexibility.

A more common problem arises with telephone and television cables that are run from the street to your house. When these have to go under concrete, the easiest and most common thing to do is to dig them directly in the dirt. The concrete is then poured over them. If this is done, you should pray that nothing ever goes wrong with either of the cables. The methods of repair are not only expensive but unsightly, unless you don't mind the sight of wires running up the outside of your house.

Have pieces of plastic conduit put under the concrete and run the telephone and TV cables in the conduit. If a cable ever needs replacing, it's then just a matter of pulling a new one through the conduit. Have an extra wire put in the conduit so that it can be used to pull in any new cable that's ever needed.

Be sure that the ends of the conduits come up through the concrete. In one house, the builder instructed the subcontractor to put the telephone and TV cables in conduit under the RV pad. This was done but the end of the conduit wasn't brought up to the surface of the RV pad. Now that the pad is poured, the two cables come up through the concrete just as if there were no conduit underneath.

Electrical power is already run in conduits so those wires should pose no problem.

RV Pads

An RV pad is another item that needs to be considered. Some subdivisions do not allow them. In others, local custom may frown on the idea of storing an RV on a residential lot. If, however, you want and can have an RV pad, you'll have to be sure that the lot is wide enough for both it and the house. Beyond that, when the builder is working up the cost for the house, be sure there's enough money for concrete to pour a useful driveway for the RV pad. Some driveways / RV pads are too narrow and simply cannot be used by RVs.

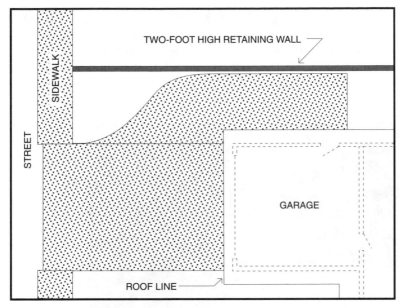

An RV pad that isn't. With the roof overhang, even the smallest RV couldn't get in there.

Decks and Patios

If you're planning a patio, make it as large as you want initially. Don't put in a small one with the intention of expanding it later because, when adding concrete to make it bigger, it is virtually impossible to make it look as if it had been poured all in one piece. It will always look like an afterthought.

Decks have similar problems. It's difficult to add on and keep anything like an aesthetically appealing appearance. So unless it's laid out for possible later expansion, make it big enough initially.

In considering whether to have a deck or patio, remember that patios are much less expensive than decks by a ratio of about 1:6 and that decks will become even more expensive as the price of lumber increases.

If you have gas in your home, don't forget to have it stubbed to the deck or patio area for your BBQ. Even if you don't have a gas BBQ, the guy that's looking at the house when you're ready to sell may find it a very positive feature. Make sure the deck isn't built over the gas stub as I've seen done a couple of times.

Watch the location of the gas stub as well as the electrical outlet on the back of your house. The gas stub is definitely for the BBQ and one of the electrical outlet's uses will be to power the BBQ spit motor. They should be close to each other and close to where the BBQ is likely to be used. This seems obvious, but more than one house was seen in which the two were are least 20 feet apart.

Gas stub and power outlets together on the patio—as they should be.

If you are having a deck built, have the decking put down with deck screws rather than nails. Screws allow you to remove a board and put it back down but, when nails are used, you have to buy new lumber and try to get a finish on it like the rest of the deck. Screws are also less likely to pull up when boards start to warp.

Electrical

Lights and Lap Siding

With lap siding or clapboard, each successive board overlaps the one below it. However, when decorative outside lights are mounted on it, there is a problem. The siding slopes, therefore the fixtures slope.

Better builders will fashion a small vertical surface on which to mount the lights and then install the siding around it. Insist on this, if you are getting lap siding—to do otherwise is just cheap.

House Numbers

You'll need lights on the front of your house to help guests find you and to see their way to the front door. These lights should be chosen for their functional use of illumination as well as their aesthetic appearance. The lights should also light your house numbers so people can find the house at night.

They may be either decorative lights which add to the decor of the house or they may be recessed can lights mounted in boxed-in soffits.

At the urging of fire and police departments, new homes in some communities use black numbers mounted on white back-lighted plastic. These units, about 5×10 inches, are equipped with light sensors which turn them on at night.

These unattractive all-houses-alike plastic numbers are not among our suggested solutions for lighting house numbers. It's a simple matter to select fixtures that shine on the house and then to

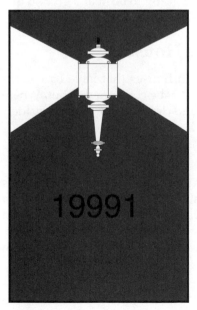
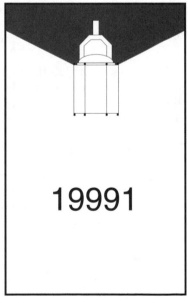

Unless you go into hibernation at night, get lights that illuminate the house numbers.

put the numbers in the light. Be sure to have this done. In over 40 percent of houses surveyed the house numbers were not lighted.

Outside Outlets

In 1993 the National Electrical Code changed the number of required outside outlets from one to two. With this change, you won't have to run an extension cord to the back of the house to get power for your Christmas lights or electric lawn mower.

If you have an RV pad, you'll want at least one out there. If you plan to have an outside barbecue with a spit or use an environmentally desirable electric barbecue starter, be sure the outlet in the back of the house is close to where the barbecue will most likely be located.

These outside outlets must be GFCI protected. It's probably a good idea for them to have their own 15 ampere or 20 ampere circuit.

If you're planning for a spa in the backyard, you'll need a separate circuit for it. According to one interpretation of the code, the outlet for the spa doesn't have to be GFCI protected. However, it's a good idea to do it. Anytime you're going to be near a ground or a water pipe and power at the same time, you can't be too careful.

Under the House

In houses with slab floors and in houses that have basements under them, there is no crawl space to be concerned about. Where there is a crawl space, both access and drainage are major considerations.

Crawl-Space Room

If the design of your house calls for a crawl space, be sure there's enough room to crawl, not just slither. The floor beams sit even with the top of the foundation wall but heating ducts and drain pipes go under the beams, thus cutting into the available space. Heating ducts are large and wrapped with insulation, making them even larger.

If your builder is going to use manufactured joists, there is another consideration. In an effort to keep the houses from sitting too high off of the ground, builders will lower the foundations to allow room for the manufactured joists. You need to be sure that the foundation is raised high enough above ground level so that there is room under the house for the joists and well-insulated heating ducts and still leave room to get around under there.

Crawl-Space Drainage

Codes don't require drain lines around a house with a crawl space. To handle a potential water problem, the builder grades the dirt under the house to a low spot and installs either a drain or a sump pump.

The dirt under the house is then covered with a heavy plastic polyethylene sheet (Visqueen). Vents around the foundation make it so that the space above the plastic is kept dry, even though the ground underneath may be wet. This keeps the wood under the house from getting dry rot, a fungus that rots wood when it's in a moist environment.

In most places, water doesn't collect under a house and builders and code inspectors get pretty sloppy about what they do and what they allow. When a house is built on a lot with a ground water problem, this sloppiness can turn into a real problem.

Consider this example:

Construction on a new house was started when the soil was still saturated from the winter rains. The dirt inside the foundation was sloped, more or less, toward one corner of the house. In actuality, the dirt was really just mud that was pushed around with a bulldozer to get something resembling a slope.

At the low corner a drain pipe was put just under the footing but was several inches above the dirt it was supposed to be draining The plastic sheet was laid down over the mud and water and the house was built.

Ground water seeped through the foundation wall, oozed up from below, and collected at the low corner of the crawl space under the house, below the level of the drain. The plastic sheet sank to the bottom of the water and the crawl space turned into a lake.

The problem should never have occurred. The drain pipe, according to the code, should have been at the lowest spot under the house. It wasn't. *Caveat Emptor.*

Where's the Sump Pump?

To solve a problem with water under a house, sump pumps are sometimes needed to keep the water from collecting. But in a neighbor's house, the pump was installed right under the master bedroom. Now in the rainy season our neighbors have to turn the sump pump off so they can sleep at night and then turn it on the next morning—and hope that the crawl space keeps dry enough to prevent dry rot.

The builder wasn't thinking of the home user when the work for the drainage under the house was being done. Vigilance is necessary. If a sump pump is needed, be sure it's located away from the bedroom area of the house.

Attic

Codes require that an attic be well-ventilated. Roof vents are used to provide the means for hot summer air to escape and, in so doing, keep the house cooler. When air conditioning is used, there's another consideration—the heating/cooling ducts, though well-insulated, must pass through the attic with its trapped summer heat. If the attic is not well-ventilated, the air in the duct will be heated, making it less comfortable and more expensive to keep cool.

Part V

Appendices

The two appendices here deal with deed restrictions (CC&Rs) and building codes. There brevity should not be misleading. Under the right circumstances ignorance of either of them can cost you money or have an unwanted and unexpected impact on the design or aesthetics of your new home.

Deed Restrictions
(a.k.a. CC&Rs)

When a builder bought a lot in an older Pacific Northwest neighborhood where the CC&Rs require cedar shake roofs at least one house had already been built in the subdivision using Woodruf (discussed in Chapter 23) which is frequently an acceptable alternative to shake. However, when the fellow across the street heard that Woodruf was planned for the new house he let it be known loud and clear that he'd go to court if the roof were not cedar shake.

Not wanting to have the property tied up in time consuming litigation, the builder caved in and built the house with cedar shake. Be careful, it could happen to you! If you don't want to get caught in a similar fix, read the CC&Rs **before** you make **any** commitment.

"Covenants, Conditions, and Restrictions," or deed restrictions, are legal conditions that are tied to the deed and tell the owners many things that may and may not be done with their property.

Purpose

CC&Rs are used to provide for continuing control over what people can do on lots in a subdivision. CC&Rs are generally beneficial. They are a legal means to protect you from the boor who doesn't care what his house or lot looks like and makes the whole neighborhood look seedy, thus ruining both your way of life and the value of your property.

Be particularly careful about what real estate agents may tell you about the CC&Rs that apply to a piece of property you may be looking at. They may have read them but probably not. I had one agent insist that there would be no problem enforcing the CC&Rs even though there was no homeowner's association. After all, the CC&Rs were the rules and why would you need the expense of an association to make people follow the rules?

It turned out this agent worked for the real estate company whose president/owner was also the subdivider. So never mind the facts, just sell, sell, sell. Caveat Emptor

What CC&Rs Mean to You

People who have lived with CC&Rs that provide for controls in the subdivision after the subdivider has gone will almost always prefer to buy in another subdivision with the same kind of CC&Rs. While they make the neighborhood a better place to live for most owners, there are a few to whom the CC&Rs are a continual source of irritation and frustration. These are the people who have their own ideas about what they are going to do on their property. It is important that you know about what's in this legal document before you make any decisions about a piece of property; you may want to find a place with tighter controls or you may not want to live with what's there already.

The title company will require that you sign a statement to the effect that you have read the CC&Rs before they will issue you a deed for your new home. Don't let it get that far. BEFORE YOU PUT A PENNY DOWN ON A PIECE OF PROPERTY READ THE CC&Rs. At first they may seem like just another piece of red tape but that document is legally binding whether you read it or not.

My wife was an officer on the board of directors of a large, strong homeowners' association and she'll tell you she couldn't believe the number of people who were unable to understand how someone could make them take down their TV antenna, repaint their house, or water their shrubbery. It's their property and no one can tell them what they can and cannot do on it, right? Absolutely wrong. If it's in the CC&Rs, it's binding. There may be some surprises although usually not of such a substantial nature as to make you want to back off. The real estate agent or seller with whom you are dealing can get a copy for you.

The term CC&R may not be an exact abbreviation for many of these documents. I have one titled "DECLARATION OF COV- ENANTS, CONDITIONS, RESTRICTIONS, ASSESSMENTS, CHARGES, SERVITUDES, LIENS, RESERVATIONS, and EASE- MENTS!" However, when you say CC&Rs people will know what you mean.

Here's a list of the most commonly seen items in some 25 differ- ent CC&Rs: easements, occupant age, street parking, driveway park- ing, home businesses, initial landscaping, antennas of any kind, nuisances, maintenance, mailboxes and newspaper receptacles, roof- ing materials, buildings, animals, trash and refuse, blocking of views, temporary buildings, sunshades, awnings, basketball back- stops, flagpoles, lights, drilling and mining, party walls and fences, clothes lines, wood piles, and outside storage areas.

CC&Rs come in all flavors and sizes. Some are only two or three pages long and others may be over 60. They are written to cover many things. The first thing they all do is to protect the land devel- oper or subdivider for as long as he has an interest in the subdivi- sion. This protection ensures that someone who buys a lot does nothing that may decrease the value of the unsold lots.

When the lots are all sold and the subdivider is gone, what then? That depends on what the CC&Rs say. The simplest ones say noth- ing. Others include the formation of a homeowners' association that can go on forever.

For the simplest CC&Rs, any benefits that may accrue to the homeowners are coincidental. The subdivider maintains architec- tural control over what may be put on a lot, but it applies only until he sells the last lot and is out of there. There may be statements in these CC&Rs about the maintenance of drainage easements, for example, but no means to enforce them after the subdivider is gone, except in a court of law. These kinds of CC&Rs are the easy way out for the land developer, but beware. When you are buying a house in one of these subdivisions, they offer little protection once the developer is gone.

Even without a homeowners' association, CC&Rs can serve a useful purpose. If someone is about to do something that is not in keeping with the CC&Rs, an owner can threaten to sue the offender. Simply the threat of the suit will often be enough to bring the of- fender into line. However, unless the offense is really gross or the suing homeowner is really sensitive, this will not usually happen because it would not be worth the cost and hassle. If it exists, the homeowners' association is the better vehicle to insist on enforcement.

CC&Rs can be amended or even deleted. How this is done should be in the CC&Rs themselves. Commonly it takes a two-thirds or three-quarters majority of the lot owners although in one CC&R this was a simple majority with the proviso that the Declarant (a.k.a. the developer) had the right of veto over any amendment.

Homeowners' Associations

Purpose

When a new subdivision is planned, there is frequently something associated with the subdivision that is unique to it and that will require maintenance over the years. An entrance that is signed, painted, and landscaped will require attention and the homeowners' association is the appropriate organization to take the responsibility to have this done. There may be streets, common recreational areas, swimming pools, parks, RV parks, and even golf courses that are owned and maintained by the association.

A primary purpose of a homeowners' association is to provide for the upkeep of these areas that are owned in common by the homeowners (the common areas).

But it doesn't stop there. The developer, in an effort to make the subdivision unique, may have adopted a theme for the subdivision. He will write the CC&Rs to be sure that everyone who buys and builds will maintain that theme.

Organization

Homeowners' associations are usually nonprofit corporations. Some CC&Rs explain in detail the makeup of the board of directors. Others say little, putting all the detail into the Articles of Incorporation, a document that is filed with the state when the association is formed. If the CC&Rs don't say just how the board is to be elected and how long the developer is to control it, you should get a copy of the Articles of Incorporation. Again, the seller or his agent should be able to get these for you from the land developer. Eventually you may want to get a copy of the bylaws for the association. This is not a legally recorded document, but it can tell you about the internal organization and operation of the association.

There is always a certain amount of administrative or management effort involved in running the association: collecting dues, answering questions, writing letters to recalcitrant owners, arranging for maintenance work, getting legal help, etc. There are organizations who specialize in this kind of administrative work. You can check to see just what arrangements have been made by the homeowners' association in the subdivision that you are considering, both for association management and for the legal counsel that any association should have.

Homeowners' associations are usually organized to be governed by a board of directors who are elected by the homeowners. Until the developer no longer has an interest in the subdivision, he usually maintains control of the board of directors, frequently by assigning himself more votes than he gives the homeowners. The exact time when control of the board of directors is turned over to the homeowners may be when the developer has sold all or most of the lots, on a predetermined date, or some combination of these.

Homeowners' associations generally have the responsibility to do those things needed to maintain the attractiveness of the subdivision and, in so doing, they can go a long way toward maintaining the aesthetics of the neighborhood and the value of the property. The association usually has, as a subsidiary group, an architectural review committee (or, variously, architectural committee or architectural control committee). With a strong set of CC&Rs, the association, through the architectural review committee, can keep a tight rein on anything that could damage property values. The architectural review committee is usually given the responsibility to promulgate a set of "architectural guidelines" about many of the details that require review. These also detail the process for getting the requisite prior approval for virtually anything you may want to do on the outside of your house.

Costs

All this does not come free. Homeowner association dues may run from less than $20 to upward of $200 per month. If this financial burden seems too onerous for the benefits that you would receive, then it would better to find a subdivision where the CC&Rs make no provision for an ongoing homeowners' association. (As noted later, you may find you have two homeowners' associations to support. So be sure you know all the facts before you invest.)

Besides the monthly dues to the homeowners' associations, you may also be subject to special assessments when there is something "special" that has to be done, such as repaving streets or completely renovating the swimming pool.

Enforcement

What happens if you don't pay your dues or assessments? The homeowners' association can file a lien against your property and either foreclose to get their money or get it, with interest, from the escrow company when you sell the house.

You will find that CC&Rs provide for penalties when the homeowner does not comply with the CC&Rs themselves or with the guidelines issued by the architectural review committee. The penalties frequently involve allowing the homeowners' association to correct the deficiency and assess a lien against the property to cover the cost of the work along with administrative and legal fees. It is then handled just as with unpaid dues, i.e., either by foreclosure or by taking the money from escrow when the house is sold (or the owner can pay the costs to the association at any time).

All the legal terms in the world won't mean a thing unless they are enforced. As long as the subdivider is around, he will take some interest in protecting his investment. Sometimes he may need some prodding when a property owner does something that definitely should not be done, but as long as the developer is looking out for his own self-interests you have some modicum of assurance that the CC&Rs will be enforced.

Don't be too sanguine about this, however. When we moved into our new house in Salem, the subdivider still had architectural control, including colors of houses. He paid little attention to this responsibility. For a new house on our block, the builder chose a ghastly shade of green. The subdivider's attitude was, "If it's too bad, it won't sell and the builder will have to repaint it." In this case the subdivider was right. It didn't sell and it was repainted. However, the subdivider's cavalier attitude did little to assure us that we had any protection from the CC&Rs at all.

With homeowners' associations things can be different, but loosely written CC&Rs may make it difficult for an association to insist on adherence to the rules. Sometimes, and this is difficult to foresee, the owners will not really be interested in keeping up the neighborhood. It is not unusual for the authority to be there, but to find that the owners themselves do not want to be constrained by the CC&Rs and, by tacit mutual agreement, do not enforce them. (You could always take the homeowners' association to court to make them do their job, but that's so costly that virtually no one will ever do it.)

Association Authority

Some people get particularly irritated by a provision found in many CC&Rs: the "right of entry." This provides that a member of the board of directors or of the architectural review committee may, from time to time, enter onto any lot in the subdivision to see if there is compliance with the requirements of the CC&Rs and/or of the architectural guidelines. And this can be done without the approval of the homeowner!

The homeowners' association and/or the architectural review committee is usually empowered to issue variances to rules or guidelines that cause a hardship in unusual cases. For example, if you have a drainage problem that cannot be fixed until dry weather comes, you could reasonably expect to be granted a variance allowing for extra time before your landscaping has to be put in.

Subassociations

In very large subdivisions, it is not unusual to find you have two sets of CC&Rs with which to deal. The primary subdivider makes up the overall set of CC&Rs for the total subdivision and then sells large parcels of the subdivision to builder/developers who, in turn, will have a set of CC&Rs that apply only to these parcels. The resulting second homeowners' association is known as a subassociation. With all of this go two sets of common areas, two boards of directors, two levels of protection, two sets of constraints and restrictions and two sets of dues. In these cases, there will usually be only one architectural review committee, from the primary subdivider's CC&Rs, that will control most of the things about which you will be concerned.

Guidelines

Once in a great while you will run across a subdivider/developer who is as concerned about your understanding of the CC&Rs as you are. When you are interested in a home or lot in his subdivision, you will get, from him or from the selling agent, a set of "guidelines" that explain in nonlegal terms what's in the CC&Rs and in the architectural guidelines and what they mean to the homeowners. With such guidelines you don't have to struggle through all the legal verbiage in the CC&Rs to find those things that are pertinent to you, the potential homeowner. They can also be comforting in that you then know, in terms you can understand, just what controls there are to maintain the appearance of your neighborhood and the value of your property.

Look for these guidelines and, when you are lucky enough to find them, you will be pleasantly surprised to find that there is, indeed, a developer who is seriously thinking about the importance of the CC&Rs and what they should mean to you.

Building and Energy Codes

Building Codes

Building codes, as they may affect your new home, are in a real state of change. There are model codes that are used entirely or in part for state codes. These state codes, in turn, may be amended by local governments. The result is that even from one locality to another in the same state there will be differing requirements. And the freedom that local governments have to amend or adopt their own codes differs from state to state.

There are model codes for each of these areas:

- Building
- Plumbing
- Mechanical (heating, air conditioning, etc.)
- Electrical
- Energy

In the early 1970s a group was formed whose purpose was to come up with a simplified set of codes dedicated specifically to detached one- and two-family dwellings. This group, The Council of American Building Officials (CABO, pronounced KAY-BO), publishes its own model code that incorporates relevant parts of the other codes. It is designed to be used as the model for state and local codes and the expectation is that by the end of the century this will lead to codes being much more alike from one place to another.

All CABO codes are revised periodically, usually every three years. Each state or local code body, in turn, takes the model codes and makes their own set of amendments to them. Thus every state or local jurisdiction has codes that may be similar to each other but differ in some respects. From the time a model code is changed until it goes through the inevitable red tape to become a part of a state or local code may take as long as five years.

Sound confusing? It is. Builders and specialists have specific codes that apply to what they do. Since the codes are revised from time to time, it's easy for the electrician, say, to get confused about what's code and what's simply local practice. And this does happen.

Energy Codes

Energy conservation has reached such importance in new construction that a new group of codes has evolved to deal with it. Here are couple of technical terms that should be mentioned.

Insulation Values

By code floors, walls, and ceilings of all new construction must be insulated. There are two terms that describe the effectiveness of insulation: R-value and U-value. They are reciprocals, i.e., R=1/U and U=1/R. R-values are used to describe the insulating quantities of materials used under floors and in walls and ceilings. U-values are used to describe the heat loss through windows, skylights, and doors. Actual building heat-loss calculations are performed by summing up the U-values of the various contributors to heat loss.

As the price of lumber goes up, there is a continuing search for alternative methods of construction. Two of these include concrete walls and metal framing members. There is a need to be careful of these when energy conservation is a consideration. Concrete offers a certain amount of insulation but, as with any other solid material, it is minimal. Since there is not a wall space that can be used, all the insulation must be placed on the inside and/or the outside of the concrete wall. When this is not handed carefully, the insulating qualities of the house will be compromised. Styrofoam sheets over the concrete are the usual solution but these then present problems in how to affix the exterior or interior siding materials. Be sure any builder who proposes using concrete walls knows what he is doing concerning appropriate insulation. If your state has energy code requirements, you may be able to get helpful advice from your building inspectors or the state code agency.

Metal framed buildings have a different problem. With normal 6" walls, the five and a half inches between inner and outer walls are filled with insulating material. But steel studs (and roofing members in vaulted ceilings) are not insulating materials and will conduct heat. This heat loss is several times greater than with wood and will lower the R value of the overall structure. Again Styrofoam sheets over the outside walls are used to compensate. And, again, be sure the builder knows what he's doing and has taken the appropriate steps to maintain the overall R value of the structure.

When these precautions are not taken, concrete walls and metal framed buildings are backward steps in the evolution toward better insulated, energy-saving homes.

Tightness

A major source of heat loss is through holes in walls and around doors and windows. To minimize this loss, houses are made tighter and tighter as time goes on. When buildings are made too tight, there is little air exchange with the outside and people in the building breathe the same air over and over. Eventually this could lead to nausea, headaches, and other complaints, particularly if there are contaminants present.

It's also a major headache for energy conservation specialists. On the one hand we don't want to lose heat and on the other hand, if we don't change the air in the house, people may get sick. There's no good way to measure when the air gets unhealthy. The solutions to this dilemma are based on experience and observations, not on hard data such as the amount of carbon dioxide or contaminants in the air. In other words, the whole process is very imprecise.

One approach is to ignore it and assume that there are enough ways for fresh air to get into the house. After all, every time you open a door or a window you get fresh air.

A second approach is to dictate the use of windows that are deliberately made with holes around them and then to use an exhaust fan to force the old air out and pull fresh air in around the windows.

A third approach is to use an air-to-air heat exchanger. A fan in this unit pulls air from the house and exhausts it and at the same time draws outside air into the house through another part of the exchanger. The unit takes heat out of the air that is being exhausted and uses this heat to warm the incoming air. These units typically recover about 35 percent of the energy that was used to heat the air initially; the other 65 percent is gone. (But that 35 percent is a whole lot more than the nothing that is salvaged with the vented-window approach.)

A further complication caused by tight houses is getting the air that exhaust fans are supposed to get rid of. Clothes dryers, kitchen exhaust fans, and the smaller bath and laundry fans can do their jobs only if there is a source of air to replace that which they are trying to exhaust. When houses are tight that air just isn't there. The whole area is very muddy with inevitable compromises among questions of energy savings, air quality, and the effectiveness of exhaust fans.

The State Codes

Individual states attack the energy conservation problem in widely varying ways. Your state may not have any specific energy conservation codes and may depend upon the insulation requirements in building codes to take care of the needs. Since CABO has now adopted a model energy code, the expectation is that more and more states will adopt that code or some version of it. The underlying idea is two-fold: 1) energy savings in general and 2) by spending money to build houses that are more energy efficient, the homeowner will save money in the long run through lower energy costs.

Depending on the state, there may be several different requirements for a state, one for each of several geographical areas.

Code Enforcement

Depending on where you live, code enforcement is a haphazard thing at best. Many states have no codes at all, leaving it up to the local jurisdictions. Others have statewide codes but allow local jurisdictions to modify them. Still others have statewide codes which cannot be changed by local government.

But in every case, it's the local jurisdiction that has the responsibility for enforcement. Which gets back to how much money a town or county may be willing to spend for code enforcement. These departments of local government are generally self sufficient insofar as the money that comes from building permits pays for the officers. In practice, the number and abilities of code officers reflect the tradeoff between the quality of the enforcement and the size of building permit fees. Needless to say, there are never enough good officers to go around. Code enforcement is never perfect and how close it comes varies widely even between adjacent towns in the same county.

A word of understanding and caution regarding building officials and inspectors. These are governmental positions and a certain amount of "politics" is involved in many of the decisions they make. In most areas the building industry is very well organized and can bring considerable pressure to bear on building officials and inspectors. The codes, unlike statutes, have not had the benefit of years of judicial decisions backing up what's written. Building officials and inspectors have to make judgments about how far to go in enforcing the letter of the code.

State laws differ, of course, but note that building officials and inspectors are liable for their decisions and are occasionally sued for not doing their job right. Generally you will find them to be cooperative.

Some codes require that a device perform a function that is impractical to measure at a building site. Then the builder and the inspector must fall back on the manufacturer's ratings and assume that the device is okay. But when the installation is not the same as when performance was measured by the manufacturer, it may not do its job when installed.

In some cases, wall insulation for example, poor installation is obvious. In others, such as bathroom exhaust fans, most installations actually render the device useless. But the inspector has no way to measure the performance so he simply checks that the fans runs when it's switched on and that's it.

Other places where the inspector may not perform his job as you may think he should are where the codes call for judgments. The word "accessible" is one that allows the inspector to be as strict or lenient as he wishes. Unless he is in the mood to punish a builder (and that happens, too), he is much more likely to let something go as accessible that most of us would not. Thus we find stop-and-waste valves hidden behind 9 inches of insulation at the far end of an inaccessible crawl space and we find whirlpool tub pumps behind access holes where the plumber would need arms like an octopus to reach them.

And there are the cases where the code may clearly spell out a criterion and then make it all vague by saying "or in accord with the manufacturer's specifications." Since the device in question probably hasn't been installed at the time of the inspection, how does the inspector know whose specifications to follow? He may just ignore the whole thing as long as something is in place—he doesn't have much choice.

And what can you, as the buyer of a new home, do to protect yourself? Probably not much. For areas that you know are likely to be given just lip service you can be very specific in your requirements. For the others you just have to go along with the code enforcement officer's okay and try to catch any gross oversights yourself.

Glossary

This listing is intended to help you understand what you're reading about in this and other books dealing with houses. It is not all-encompassing; many terms builders use are omitted because they don't generally involve the home buyer.

3-way switch—Uses two switches that permit a light or outlet to be turned on or off from either switch. The term "3-way" refers to the three ways which the switches can be set: 1) OFF, 2) ON from switch A, or 3) ON from switch B.

4-way switch—Similar to a 3-way switch except that there are three switches instead of two. The light can be turned on or off from any of the three.

A—Ampere. A measure of electric current, used with a number (15A, for example) to indicate the capacity of a circuit, fuse, or circuit breaker.

ac—In today's houses this usually means the air conditioning unit. It can also mean alternating current to differentiate it from direct current.

all-in-one—Type of mortgage loan that includes the construction loan and the permanent home loan in one package. Used with custom or build-to-order houses. Also called a "wrap-around" mortgage.

Amp—(*See* A.)

backerboard—A special rigid board used under tile.

backfill—Dirt that's pushed back against the house foundation after the house is finished.

backsplash—Area just behind and above a countertop, usually covered to prevent water from splashing on the wallboard.

baseboard—Piece of material used to hide the joint where the floor meets the wall; may be wood, plastic, or rubber.

base molding—Molding used as a baseboard.

batten—Narrow piece of material used on the outside of a house to cover joints in walls.

bearing wall—Wall that supports a ceiling joist, floor joist, or roof. May be an outside wall or an inside partition.

bid out—Refers to the process builders use to estimate the cost of a house before it's built.

board and batten—Siding made up of vertical boards the joints of which are covered with batten slats.

board foot—Amount of lumber in a piece 1" thick by 12" wide by 1 foot long. A piece of 1" × 12" that is 8 feet long has 8 board feet. A 2" × 6" that is 8 feet long also has 8 board feet.

boot—Piece of formed sheet metal used to interconnect a heating/cooling duct and a register.

BR—Bedroom. Could also mean brick.

branch circuit—Electrical circuit with its own circuit breaker in the service panel.

brick veneer—Brick facing on an exterior wall or fireplace.

building codes—Generic term relating to codes for buildings. (*see* codes.)

building inspector—A government employee who inspects buildings to determine if they are code compliant.

building official—A government official with the responsibility for generating codes and/or enforcing them. May supervise building inspectors.

bulllnose—The rounded outside edge of a tile counter.

butt—Describes how the ends of two boards or the edges of sheet materials meet so that their ends or edges touch in a continuous line. Also describes the end of a board.

CABO—Council of American Building Officials. (*See* Appendix B.)

can light—An incandescent light in a metal can. The can is flush with the surface in which it is mounted. The light itself may be recessed.

Casablanca fan—(*See* ceiling fan.)

casement window—Windows that hinge in or out for opening.

casing—Trim for a window, door, or opening.

caulking—Refers to the compound used to make joints weatherproof and waterproof or simply to make them smooth.

CC&Rs—Covenants, Conditions, and Restrictions. Appended to deeds. (*See* Appendix A.)

ceiling board—Like wallboard (which *see*) but more rigid so that it won't sag when installed horizontally in a ceiling.

ceiling fan—Large ceiling-mounted fan. Also known as a paddle fan or Casablanca fan. Although Casablanca is the name of one manufacturer of ceiling fans, the name is also used generically.

center mullion—In some cabinets, the vertical piece of wood on the front that divides the opening into two parts.

chase—A boxed-in shaft, usually vertical, through which pass various pipes, drains, ducts, and flues.

chipboard—A board made of wood chips that are glued together under pressure

CI—Cast iron.

clapboard—A siding, usually cedar, made of overlapping horizontal boards. Largely replaced by lap siding.

cleanout—Part of a fireplace from which ashes may be removed from the fireplace.

codes—A group of legal documents that define many building parameters. (*See* Appendix B.)

comps—Comparables; describes the process of comparing the values of different pieces of property. Used by appraisers as a major tool in determining property values.

conduit—A pipe or tube through which wires or smaller pipes can be run.

construction loan—A short-term loan made to a builder to finance the construction of a building. Money is loaned only to cover the builder's costs for materials and labor as it is spent.

cop—Copper, as in copper pipe.

crawl space—Space between the bottom floor of a house and the ground under it.

crown—Piece of molding around the top of a room.

D—Dryer.

deck—1) the flat area around a bath tub, *see* also surround; 2) with post and beam construction, the flat wooden surface on top of the beams; 3) a wooden platform on the exterior of houses used for outdoor living.

diag—Diagonal.

dim—Dimension.

dimensional lumber—Single pieces of lumber that are sawed to standard dimensions as opposed to that which is "manufactured."

direct vent—A type of gas fireplace that vents the burned gas directly to the outside of the house.

dn—Down, as in stairs.

door light—The small vertical window(s) on one or both sides of an entry door.

double-hung—Two-window sash where each window slides up and down.

DR—Dining room.

draw—A payment made to builder against a construction loan.

drawer boxes—The boxes in cabinet drawers to which are attached the facing piece drawer slides.

dry—Dryer.

drywall—(*See* wallboard).

duct—A pipe, usually metal, used to carry heating or cooling air.

ductless—a kitchen exhaust scheme where the fumes, steam, heat, etc., are run through a charcoal filter and blown back into the room.

duplex outlet—The standard electrical outlet in a home. (Two separate plugs can be plugged in, hence the term "duplex.")

DW—Dishwasher.

eaves—The part of the roof that overhangs the outside wall of a building.

el—Elevation.

elevation—A drawing showing the straight-on exterior view of a building.

empty nester—Refers to an older couple whose children are gone, leaving them with an empty house.

ent—entrance.

entry level—(*See* starter house.)

European-style—Describes a style of cabinet without a face frame, also called "box" cabinets. Also used to describe a particular type of hinge needed for this style of cabinet. The hinge is not visible from outside the cabinet.

exfiltration—Leakage of air from the inside to the outside of a building.

ext—Exterior.

face frame—A term used to identify the ring or frame of wood that goes around the front opening of western- or face-frame style cabinets.

fascia—Board covering ends of the roof rafters.

fenestration—Refers to windows and the way they are arranged in a building.

fl—Floor.

flip switch—Electrical switch that is operated by moving the control up/down rather than pushing or tapping on it (as differentiated from a rocker switch, which *see*.)

footing—The base on which a building's foundation sits.

FP—Fireplace.

gable—Describes a triangular section of a wall that extends up from the level of the eaves to a roof peak. The roof line is straight, not curved or broken.

gar—Garage.

GFCI—Ground-fault circuit-interrupter.

grout—A kind of mortar used to fill between tiles, marble, or stone.

gl—Glass. Used on drawings to show sliding glass doors.

glazed—From the verb glaze meaning to fit a window with glass panes. Double glazed is used to describe a window with double panes.

gyp board—(*See* wallboard.)

gyp rock—(*See* wallboard.)

gypsum—A white mineral. (*See* wallboard.)

hardboard—A material made from wood fibers that may be made into sheets or pieces to simulate wood.

heat pump—Part of a heating and air-conditioning (HVAC) system. Pumps heat into or out of the house.

hollow-core door—A door whose interior is mostly empty.

hose bibb—An outside faucet to which a hose may be connected.

hugger—A type of ceiling fan that uses a minimum of vertical space. Used with low ceilings.

HVAC—Heating, ventilation, and air conditioning.

ICBO—International Council of Building Officials.

in—Inch or inches.

infiltration—Leakage of air from the outside to the inside of a building.

joist—A horizontal beam supported by a bearing wall. Usually used to support floors and/or ceilings.

kiln-dried lumber—Lumber that has been dried in a kiln or oven. Differentiated from lumber that may be wet when used and then dries after the house has been erected.

kit—Kitchen.

laminate—A plastic surfacing material used on countertops and, rarely, in tub and shower surrounds. Popularly known by the trademarked name Formica.

lap siding—Boards of wood or composite material used horizontally on the outside of houses in which the bottom of one piece is made to overlap the top of the piece below it.

light—(*See* door light.)

lin—Linen closet.

lndry—Laundry.

LR—Living room.

MBR—Master bedroom.

mechanical—Dealing with heating, ventilating, and air conditioning as in "Uniform Mechanical Code" and "mechanical subcontractor." (In some parts of the country, mechanical is used to refer to electrical and plumbing as well as the items covered in the Uniform Mechanical Code.)

Mello-Roos—Refers to special assessments districts in California formed to pay for streets, sewers, etc., in new subdivisions. Can add significantly to the tax bill.

miter—Used to describe how two boards meeting at right angles are cut so that cut ends do not show. Miters are usually 45-degree cuts.

MLS—(*See* multiple listing service.)

modular house—a made-to-order factory-built house that is not constrained to pre-determined designs.

molding or moulding—Strips of wood, usually decorative, used as trim. May be painted or stained and finished.

move-up house—The next step above a starter house. A house for a growing family.

multiple listing service—A service provided by the local real estate agents that lists all properties for sale in an area along with many of their more pertinent features.

NAR—National Association of Realtors.

NAHB—National Association of Home Builders.

no duct—(*See* ductless.)

nosing—The front edge of a stair tread that extends or noses over the riser.

no vent—(*See* ductless.)

NV—No vent. (*See* ductless.)

O—Built-in oven.

OC—(*See* on center.)

on center—Used to specify distance from the center of one piece of material to another. Most frequently used to describe the distance between studs or joists.

OD—Outside dimensions.

orange peel—Refers to a wallboard finish that resembles the texture of an orange peel.

OSHA—The Occupational Safety and Health Administration. A federal agency.

P—Pantry.

paddle fan—(*See* ceiling fan.)

party wall—A wall or fence that sits on a property line and is a common responsibility of both parties or land owners.

particle board—Similar to chipboard except that the pieces are smaller (particles rather than chips). Has a higher density than chipboard.

pass through—An opening in a wall to pass dishes through. Usually between kitchen and breakfast nook or dining room.

pitch—May refer to the sticky stuff that exudes from lumber but more likely has to do with the slope or pitch of the roof. A

4/12 pitch, for example, means that there is 4" of vertical rise for each 12" of horizontal run.

plasterboard—(*See* wallboard.)

plate—The bottom or top piece of a wall. Studs sit on the bottom plate and are capped by the top plate.

pocket door—An interior door that slides into a pocket in a wall.

pot mark—Marks made on tile surfaces by aluminum and other metal pots and pans.

production house—one made in a factory-like production-like envivornment where it is one of many essentially identical houses. Usually in tract subdivisions.

PWR—Powder room or half-bath (not power).

R—Range, when used in kitchens.

r—Radius.

R-value—Resistance to heat flow. (*see* Appendix B.)

recirculating—*See* ductless. May also refer to a recirculating hot-water system that provides immediate hot water to faucets in the house.

ref—Refrigerator.

resilient channel—A channel (ususally steel) run across wall studs under wallboard to reduce sound transmission through the studs.

riser—The vertical boards between the steps of a stairway.

riser height—The vertical distance from the top of one step of a set of stairs to the top of the next.

rm—Room.

rocker switch—An electrical switch which is operated by tapping the top or bottom of the control plate (as differentiated from a flip switch, which s*ee*.)

RV pad—A concrete pad specifically intended for storing a recreational vehicle.

sawn lumber—(*See* dimensional lumber.)

sconce—Wall-mounted light fixture.

scr—Screen.

shake—A thick wood split shingle, usually cedar, used for roofs and siding.

sheet rock—(*See* wallboard.)

sidelight—(*See* door light.)

sillcock—(*See* hose bibb.) (This term is seldom used in the west.)

soffit—The underside of eaves. Also used to designate the space between cabinets and the room ceiling. May be closed (boxed-in) or open.

solid-core door—A door whose interior is filled as differentiated from a hollow-core door.

solid surface—Generic name to a family of materials used for counters, shower stalls, etc. These materials (acrylic, polyester or a mix) are the same throughout, i.e., not just on the surface.

solid-surfacing veneer—A laminate of a thin solid surface material laid on a particle board or plywood backing.

soundboard—A special board used under wallboard to reduce sound transmission through the wall.

spa—(*See* whirlpool tub.)

spec—A house built on speculation with the expectation of finding a buyer for it. Differentiated from a custom or build-to-order house for a which there is a buyer before building starts.

specs—Specifications.

SSV—(*See* solid-surfacing veneer.)

starter house—A family's first house. Usually small and low-cost.

stor—storage.

studs—The uprights used in walls. Usually 2×4s or 2×6s. Usually wood in residential construction but may be metal.

stucco—A siding material made with Portland cement that is spread on in several layers over a metallic mesh.

subcontractor—A company or person employed by the builder to do a specialized part of the house construction.

subdivider—A company or person who divides land into lots to sell or on which to build houses. Usually includes streets and utilities in the process.

subfloor—What's under the finished floor. Refers to the rough material laid across the floor joists to support the rest of the floor. Also can refer to the sheathing, plywood, or particle board, used to give a smooth surface for the finish flooring.

surround—Refers to the material that surrounds a bathtub or shower. Tub surrounds may be only the flat area around the tub (the deck) and may include a facing on the front of the tub as well as a backsplash.

T1-11—A hardboard (Masonite) exterior siding with vertical grooves to simulate boards. Comes in 4' × 8' or 4 × 10' panels.

T111—*(See* T1-11.)

T&G—*(See* tongue and groove.)

TC—*(See* terra cotta.)

terra cotta—A red low-fired tile. Used extensively for roofing in the desert southwest but now being superseded by concrete tile because of concrete's greater strength.

tongue and groove—Lumber with a small groove down one side of each board and a protruding piece (the tongue) that fits into the groove when the boards are installed.

tract—A land development area, typically with models of houses for sale.

tread—The flat part of a stair step.

tread width—The front-to-back width of a stair step exclusive of the nosing.

typ—Typical, as in a typical stud arrangement.

unvented—A type of gas fireplace or stove that doesn't vent the spent gasses outside but dumps them back into the room.

U-value—Used to quantitatively describe the amount of heat a door or window conducts between inside and outside air. (*see* Appendix B.)

vaulted ceiling—Arched ceiling.

V-Cap—A special piece of ceramic tile used at the edge of a counter to prevent liquids from running off. Name comes from underside shape which is a V.

visqueen—Polyethylene sheet used to cover the ground in a crawl space.

VT&G—Vertical tongue and groove.

W—Clothes washer.

walk-through—Joint action taken by the buyer and builder at completion of the house to find those items that have not been done properly and need to be fixed by the builder.

wallboard—Strictly speaking, refers to all wall materials that come in sheets including plywood and Masonite hardboard. We use "wallboard" here to mean sheets of compacted gypsum (a mineral) with a paper exterior. This is, by far, the most commonly used wallboard. Also called dry wall, sheet rock, gyp rock, gyp board, gypsum board, plasterboard and probably more.

WC—Water closet or toilet.

western-style—Describes a cabinet with a face frame around its front opening.

whirlpool tub—More popularly known by the trade name Jacuzzi and also called spa. A tub with jets of water flowing into it to give a whirlpool effect.

WR—Washroom.

wrap-around—(*See* all-in-one.)

X—Marks a location.

yd—Yard.

zero lot line—A house has a zero lot line when it is built with one wall on the edge of the lot. This wall is frequently a common wall with the adjacent house.

Other Reading

Books

One of the difficulties you'll find in collecting information for your new home is the plethora of books on structure and aesthetics—how to build houses and how to make them look pretty—and the paucity of books on function.

You'll find small books and large ones, some with pretty colored pictures and some that are useful. If you're attracted to the superficial—because it's easier to read and doesn't require any thinking—you'll find that kind of book, too.

The following list is by no means complete; you will find many more useful books when you visit your local bookstore or library.

Function

Books published by the National Kitchen& Bath Association are for members and, while generally for remodelling, have a wealth of information for anyone interested in designing functional kitchens and baths.

Patrick J. Galvin, *Kitchen Basics, A Training Primer for Kitchen Specialists*, 4th Edition (E. Windsor, NJ: Galvin Publications, 1995).
> Contact NKBA, 800/367-6522 (option 1). I do not agree with all of their guidelines but the depth of the discussions makes this book worthwhile.

Aesthetics

Books discussing aesthetics range from scholarly descriptions of archtectural design to personal views on what color of paint goes where.

Joanna Wissinger, "The Interior Design Handbook," (New York, New York: Henry Holt & Co., 1995)
> A combination of functional and aesthetic considerations, the interior designer's view differs somewhat from *BUILD IT RIGHT!* The room-by-room checklists will help in setting your priorities.

Structure

These how-to books are generally for the do-it-yourselfer, whether it's being your own general contractor or doing some or all of the actual construction.

John Folds and Roy Hoopes, *Building the Custom Home: How to Be Your Own General Contractor* (Dallas, TX: Taylor Publishing Co, 1990).

Jim Locke, *The Well-Built House*, Revised Edition (Boston/New York: Houghton Mifflin Co., 1992).
> From the builder's perspective with his personal likes and dislikes.

R. Dodge Watson, *Build Your Dream House For Less* (Cincinnati, OH: Betterway Books, 1995).
> Save money by being your own general contractor.

Financing and Management

Regardless of how you buy your new home, you'll need help in arranging financing and in managing the sale of your old house as well as getting the new one.

Alan and Denise Fields, *Your New House: The alert consumer's guide to buying and building a quality home.* Second Edition (Boulder, CO: Windsor Peak Press: 1996).
> A scary relating of the many things that can go wrong during the home buying and construction process.

Dian Hymer, *Starting Out :The Complete Home Buyer's Guide* (San Francisco, CA: Chronicle Books: 1997).
> A real estate agent's explanation of the home buying process including discussions of financing and sales agents. Doesn't cover buying a tract, spec, or custom house.

Staff of Kiplinger's Personal Finance Magazine, *Buying & Selling a Home*, 5th edition (Washington, DC: Kiplinger Washington Editors, Inc., 1996).
> A thorough broad view of what's involved in the buying and selling processes with emphasis on the financial aspects.

Carol Smith, *Building your Home: An Insider's Guide* (Washington, DC: Home Builder Press, 1996)
> Published by the NAHB, it covers the process of dealing with a builder to get a custom home built. Doesn't mention the competitive bid process and there is no discussion of alternate ways to get a house built.

Magazines

Books aren't the only place to get inputs. Many, many periodicals are published on a monthly and quarterly basis that have ideas
you may be able to use. These you'll find when you visit the supermarket. A word of caution. Unlike books, there are no reviewers to
rate magazines and you'll find that most periodicals are a combination of the meaty, the eye-catching, and the trivial. But don't overlook the advertising; it is a great source of ideas for materials for
your new home. Writing or simply calling the toll-free numbers
can get you all of the colorful and informative brochures you could
ever hope for.

Index

New Home Checklist

This is a revised version of the popular checklist used by readers of the first edition of *BUILT IT RIGHT!*

It provides a systematic way to check a house or a floor plan to be sure you haven't overlooked one of the hundreds of details that can come back to haunt you later.

Each of the over 300 items in the checklist is cross referenced to the page in *BUILT IT RIGHT!* where it is discussed.

Using the checklist you go through the house, system by system and room by room, checking each item. There is space at the bottom of each page for notes.

Available only from Home User Press, the checklist is $4.95 including shipping and handling.

send check to: or call toll free:
Home User Press   **1-800-530-5105**
1939 Woodhaven Street NW   **Visa/MC/AmEx**
Salem, OR 97304